COMPUTING PREDICTIVE ANALYTICS, BUSINESS INTELLIGENCE, AND ECONOMICS

Modeling Techniques with Startups and Incubators

Innovation Management and Computing

COMPUTING PREDICTIVE ANALYTICS, BUSINESS INTELLIGENCE, AND ECONOMICS

Modeling Techniques with Startups and Incubators

Edited by

Cyrus F. Nourani, PhD

Apple Academic Press Inc.
3333 Mistwell Crescent
Oakville, ON L6L 0A2
Canada USA

Apple Academic Press Inc.
1265 Goldenrod Circle NE
Palm Bay, Florida 32905
USA

Exclusive worldwide distribution by CRC Press, a member of Taylor & Francis Group

International Standard Book Number-13: 978-1-77188-729-8 (Hardcover)
International Standard Book Number-13: 978-0-42946-546-8 (eBook)

CIP data on file with Canada Library and Archives

Library of Congress Cataloging-in-Publication Data

Names: Nourani, Cyrus F., editor.

Title: Computing predictive analytics, business intelligence, and economics : modeling techniques with startups and incubators / editor, Cyrus F. Nourani, PhD.

Description: Oakville, ON, Canada ; Palm Bay, Florida, USA : Apple Academic Press, 2019. |
Includes bibliographical references and index. |
Identifiers: LCCN 2019017468 (print) | LCCN 2019018357 (ebook) | ISBN 9780429465468 () |
ISBN 9781771887298 (hardcover : alk. paper) | ISBN 9780429465468 (ebook)

Subjects: LCSH: Management--Statistical methods. | Business intelligence. | New business enterprises.

Classification: LCC HD30.215 (ebook) | LCC HD30.215 .C66 2019 (print) | DDC 658.1/1011--dc23

LC record available at https://lccn.loc.gov/2019017468

Apple Academic Press also publishes its books in a variety of electronic formats. Some content that appears in print may not be available in electronic format. For information about Apple Academic Press products, visit our website at **www.appleacademicpress.com** and the CRC Press website at **www.crcpress.com**

INNOVATION MANAGEMENT AND COMPUTING

Innovation is the generation and application of new ideas and skills to produce new products, processes, and services that improve economic and social prosperity. This book series covers important issues in the burgeoning field of innovation management and computing, from an overview of the field to its current and future contributions. The volumes are of value and interest to computer and cognitive scientists, economists, engineers, managers, mathematicians, programmers, and engineers.

SERIES EDITOR:

Cyrus F. Nourani, PhD
Research Professor, Simon Fraser University, British Columbia, Canada;
Academic R&D at Berlin, IMK Bonn, and Munich, Germany
E-mail: Acdmkrd@gmail.com or cyrusfn@alum.mit.edu

Current books in the series:

- Ecosystems and Technology: Idea Generation and Content Model Processing
- Computing Predictive Analytics, Business Intelligence, and Economics: Modeling Techniques with Startups and Incubators
- Affective Haptics, Neurocognition, and Emotional Recognition for Androids and Machine Learning (forthcoming)
- Innovations on Cooperative Computing and Enterprise Blockchains (forthcoming)

Tentative forthcoming volumes in the series:

- Diaphontaine Sets, Definability and Newer Developments on Computability, Languages, Sets, and Model Theory
- Economic Game Models and Predictive Analytics
- Epistemic Computing, Model Checking, Spatial Computing and Ontology Languages
- Content Processing, Intelligent Multimedia Databases, and Web Interfaces
- Agent Software Engineering, Heterogeneous Computing, and Model Transformations

CONTENTS

ABOUT THE EDITOR

Cyrus F. Nourani, PhD
Research Professor, Entrepreneur, and Independent Consultant

Dr. Cyrus F. Nourani has an international reputation in computer science, artificial intelligence, mathematics, virtual haptic computation, enterprise modeling, decision theory, data sciences, predictive analytics economic games, information technology, and management science. In recent years, he has been engaged as Research Professor at Simon Frasier University in Burnaby, British Columbia, Canada, and at the Technical University of Berlin, Germany, and he has been working on research projects in Germany, Sweden, and France. He has many years of experience in the design and implementation of computing systems. Dr. Nourani's academic experience includes faculty positions at the University of Michigan–Ann Arbor, the University of Pennsylvania, the University of Southern California, UCLA, MIT, and the University of California, Santa Barbara. He was Visiting Professor at Edith Cowan University, Perth, Australia, and Lecturer of Management Science and IT at the University of Auckland, New Zealand.

Dr. Nourani commenced his university degrees at MIT, where he became interested in algebraic semantics. That was pursued with a world-renowned category theorist at the University of California and Oxford University. Dr. Nourani's dissertation on computing models and categories proved to have pure mathematics foundations that were published from his postdoctoral times at US and European publications. He has taught AI to the Los Angeles aerospace industry and has worked in many R&D and commercial ventures. He has written and coauthored several books.

He has over 400 publications in computing science, mathematics, and management science, and he has written and edited several volumes on additional topics, such as pure mathematics; AI, EC, and IT management science; decision trees; and predictive economics game modeling. In 1987, he founded ventures for computing R&D and was a consultant for such clients as System Development Corporation (SDC), the US Air Force Space Division, and GE Aerospace. Dr. Nourani has designed and developed AI robot planning and reasoning systems at Northrop Research and Technology

Center, Palos Verdes, California. He also has comparable AI, software, and computing foundations and R&D experience at GTE Research Labs.

CONTRIBUTORS

Rajat Agarwal
Department of Management Studies, Indian Institute of Technology, Roorkee, India

Sverker Alänge
Department of Technology Management and Economics – STS, Chalmers University of Technology, Gothenburg, Sweden, E-mail: sverker.alange@chalmers.se

Henry Etzkowitz
International Triple Helix Institute, Palo Alto, USA

Johannes Fähndrich
Technical University of Berlin, DAI Labs, Berlin Germany

Kristina Fajga
Center for Entrepreneurship, Technische Universität Berlin, Hardenbergstraße 38, 10623 Berlin, Germany, E-mail: kristina.fajga@tu-berlin.de

Gopinath Ganapathy
School of Computer Science and Engineering, Bharathidasan University, Tiruchirappalli, Tamilnadu, India, E-mail: gganapathy@gmail.com

Sudhanshu Joshi
School of Management, Doon University, Uttarakhand, India

Rob Van Kranenburg
Founder of Council, Involved in NGI.eu (NGI Move and NGI Forward), Member of dyne.org and TRL, DeTao Master, Ghent, Belgium

Jan Kratzer
Chair of Entrepreneurship and Innovation Management and Managing Director, Center for Entrepreneurship, TU Berlin, Straße des 17, Juni 135, 10623 Berlin, Germany, E-mail: jan.kratzer@tu-berlin.de

Codrina Lauth
Copenhagen Business School, Digitalization Department and Grundfos Technology & Innovation, Denmark

Pankaj Madan
Faculty of Management Studies, Gurukula Kangri Vishvavidhaylaya, India

Cyrus F. Nourani
SFU, Computation Logic Labs, Burnaby, BC, Canada, Acdmkrd: AI, Berlin, Germany, E-mail: cyrusfn@alum.mit.edu

Ilona Pawełoszek
Częstochowa University of Technology, Faculty of Management, E-mail: ilona.paweloszek@wz.pcz.pl

Oliver Schulte
SFU, Computation Logic Labs, Burnaby, BC, Canada, Acdmkrd: AI Berlin; Akdmkrd.tripod.com DE, the Academia California and SFU, Burnaby, Canada, E-mail: OSchulte@cs.sfu.ca

Annika Steiber
Menlo College, Atherton, California, USA, E-mail: annika.steiber@menlo.edu

Chellammal Surianarayanan
Department of Computer Science, Bharathidasan University Constituent Arts & Science College, Tiruchirappalli, Tamil Nadu, India, E-mails: chelsrsd@rediffmail.com; chelsganesh@gmail.com

Deepa Vijay
School of Computer Science and Engineering, Bharathidasan University, Tiruchirappalli, Tamil Nadu, India, E-mail: deepa.vijay1@gmail.com

ABBREVIATIONS

AI	artificial intelligence
BI&A	business intelligence and analytics
BPD	business process diagram
BPEL	Business Process Execution Language
BPMN	Business Process Model and Notation
CMS	competency management system
COPL	CeleraOne Programming Language
DARPA	Defense Advanced Research Project
DEA	data envelopment analysis
EM	experience management
ERP	Enterprise Research Planning
FOL	first-order language
GEA	GE Appliances
GF-diagram	generalized free diagram
IBM	pervasive computing
ICT	Information and Communication Technology
IDF	inverse document frequency
JCR	Journal Citation Report
JVSV	Joint Venture Silicon Valley
KM	knowledge management
KR	knowledge representation
LAN	local area network
LSA	latent semantic analysis
MCDM	multi-criteria decision making
MSc	Master of Science
NE	Nash equilibrium
NIH	not-invented-here
OM	operation management
OR	operation research
PRS	procedural reasoning system
RDCEO	Reusable Definition of Competency or Educational Objective
SCI-expanded	Social Citation Index Expanded

SCM	software configuration management
SCO	Software Competence Ontology
SDK	software development kit
SICRO	synchronous insert-compute-respond operations
SRI	Stanford Research Institute
SSCI	Social Science Citation Index
ST	systems thinking
SVD	singular value decomposition
TAM	technology adoption model
TAT	technology adoption technique
TBL	triple bottom line
TF	term frequency
TF-IDF	term frequency-inverse document frequency
VSM	vector space model
WAN	wide area network

PREFACE

The goal of this volume is to bring together research and system designs that address the scientific basis and practical design issues that support areas ranging from intelligent business interfaces and predictive analytics to economics modeling.

In recent years, applications for management science and IT have been areas of interest for business schools and computing experts. Among the areas that are being treated are modern analytics, heterogeneous computing, business intelligence, ERP (enterprise resource planning), and decision science. Consumers have been pledging their love for data visualizations for a while now, and data exponents or models are explored for area sectors such as B2B and EC (e-commerce), e-business and the Intelligent Web, CRM (customer relationship management), and infrastructures.

The digitization implications of these many new applications are described and explored in this informative volume. The volume starts with a report of university startup incubators brought forth from Technical University Berlin's Center for Entrepreneurial Management; that is, among the major Europe university startup players during recent years. Chapter 1 presents insights into the transfer of technology from universities regarding innovative high-tech startups in Berlin, Europe's no. 1 startup hub. Additional chapters on the startup or combined big corporation startups examine brilliant compositional views on how startups can play an interim role with bigger corporations, exploring a "best of both worlds" scenario, presented by the Berkeley Research Group and a Chalmers University, Götborg, Sweden team.

The chapter authors address diverse issues in conjunction with computing predictive analytics, business intelligence, and economics, including university startup incubators, innovation in business, high-tech startups, strategic leadership, developing management systems, sustainable manufacturing and services, strategic decision trees, identifying business competencies, and more.

The volume's thematic topics include business interface design, business planning uncertainty models, big data, data reduction, and governance on enterprise modeling. The authors become specific to point out that large companies tend to lose their innovativeness with new procedures and that regulations aimed at strengthening key business processes simultaneously can stifle innovation. To improve existing products and services, the authors assert the need to simultaneously pursue different strategies. This question of ambidexterity has been on the industry agenda for many years. An empirical case presented as an example of a new approach to deal with the demands for ambidexterity in a large corporation, deploying an independent R&D unit acting like a startup within.

The volume presents newer glimpses on how the newer building blocks for digital transformation might indicate directions for middle management roles that must, in reality, handle the transformation. An example shows how a Europe IOT digital transition leader describes the depth of disruption that is caused by multiple "drivers" or "trajectories" toward a pragmatic solution to a digital transition. The author describes four main drivers of the digital transition: digitization of our analog world, the changing nature of power, the 'extra' on techno or dataism, and managing motivation. These drivers are pushing three main trajectories of digitization in our built environment—the 'smart city'; gated communities; decentralized localism—toward pragmatic smart city cybernetics.

In the chapter "Strategic Decision Trees on Impact Competitive Models," the authors summarize several years of research since they were embedded in graduate studies at Copenhagen Business School and the Western Danish critical industry areas. Ever since our publications on frontiers of decision trees and forecasting over a decade ago, the techniques for enterprise business systems planning and design with predictive models have become a focus of attention. The areas addressed are designing predictive modeling with strategic decision systems with applications to analytics, enterprise modeling, and cognitive social media business interfaces.

Planning goal decision tree satisfiability with competitive business models distinguishes impact goals to focus on goals that can be processed with MIS models. An important new area presented is on systems for identifying competencies needed to achieve a company's goals. The competencies are viewed in the framework of tasks implemented within the business processes. The complex nature of competencies requires expressive forms of descriptions regarding the multidimensional character; therefore, codification of competencies is the area predestinated to make use of ontologies. A specific system that can be useful in competitive business modeling based on ontologies is described to support process-oriented, dynamic competency management in a company. The presented approach to manage competencies is based on business process flows that can be valuable for process-based and virtual modeling, with an environment that is dynamic with frequent changes that are needed to preserve competitiveness and agility.

So far as practical or sociological glimpses of the heart of Silicon Valley high-tech entrepreneurial panorama, we have authors presenting the high-tech urbanization with no bounds of nature, counter-culture, or exurban or urban life. The panoramas are expanding into the area with Google and Apple, or Facebook, and are outgrowing the willingness of smaller cities such as Mountain View and Cupertino to accommodate their growth. Movements are into and above Berkeley, spreading across counties formerly considered as part of the San Francisco Bay Area, or to the central valleys to the east. The authors note that the Stanford Research Park is virtually invisible in Silicon Valley, while the triple helix to thrive startups must involve a high-tech university to model such innovative centers. Today, the Park hosts the headquarters of the two descendant firms of Hewlett Packard, the Skype subsidiary of Microsoft, various law firms, the StartX accelerator, and other elements of the local innovation ecosystem. The chapter discusses the sources of Silicon Valley's success and issues that have arisen due to "too much success," inducing urban or social population movements that find the area so overpriced for living that the average person cannot afford to live there and newer startups cannot afford to be there to thrive without significant investment factors that can be accommodated.

A more technical chapter presents new economics analytics models with manageable computability properties for competitive learning models with goal game tree planning. New model computing techniques for Van Neumann Morgenstern Kuhn games are developed. Furthermore, new heuristics for matrix Nash games are presented. To enhance the pure analytics contents, a chapter on random set models examines random big data computations for competitive models. Novel techniques are presented to reach goals with data models on computable decision trees. For more on analytics areas, a chapter presents new visualization techniques with analytics applications. Combining a meta-contextual logic with Morph Gentzen, a new computing logic that the author has developed since 1997 is presented. A visual forecasting analytic with examples is presented. Newer applications toward virtual reality computing are developed with smart language features carry visual structures via functions.

The chapter "Sustainable Manufacturing and Services in Industry 4.0: Research Agenda and Directions" examines a comprehensive review of cross-disciplinary literature in the domain. A noted list of 3561 articles from journals and search results are used from academic databases. The reason for that adoption over existing bibliometric techniques is to use the combination of text analysis and mining methods to formulate a latent understanding future research trends that will assist researchers in the area of manufacturing and service operations; therefore, the study will be beneficial for practitioners of the strategic and operational aspects of technology adoption in manufacturing and service supply chain decision making.

Ontology-based approaches to developing realistic designs are presented in a chapter. Successful completion of a software project purely depends on better understanding of the requirements of the clients in order to avoid ambiguity issues and improve the understanding of the requirements. An ontology-based approach is proposed for representing software requirements. Ontological concepts are developed for individual tasks. It is proposed to employ the application of ontology to similar kinds of projects so that it will evolve as domain ontology.

A generalized Min-Max local optimum dynamic programming algorithm is presented and applied to business model computing where predictive techniques can determine local optima. Systemic decisions are based on common organizational goals, and as such, business planning and resource assignments should strive to satisfy higher organizational goals. Causal loops are applied to complex management decisions. A preliminary optimal game modeling technique is presented in brief, and the example application areas to e-commerce with management simulation models are examined.

CHAPTER 1

UNIVERSITY STARTUP INCUBATION IN EUROPE'S STARTUP METROPOLIS NO. 1: THE CASE OF TECHNISCHE UNIVERSITÄT BERLIN

KRISTINA FAJGA[1] and JAN KRATZER[2]

[1]Center for Entrepreneurship, Technische Universität Berlin, Hardenbergstraße 38, 10623 Berlin, Germany, E-mail: kristina.fajga@tu-berlin.de

[2]Chair of Entrepreneurship and Innovation Management and Managing Director, Center for Entrepreneurship, TU Berlin, Straße des 17, Juni 135, 10623, Berlin, Germany, E-mail: jan.kratzer@tu-berlin.de

ABSTRACT

This chapter provides insights into the transfer of technology from universities regarding innovative high-tech startups in Berlin, Europe's no. 1 startup hub. Thereby, the case of Technische Universität Berlin (TU Berlin), a pioneer of entrepreneurship support activities in Germany, is presented. Berlin is considered as one of the most entrepreneurship-friendly cities in Europe. One of the country's oldest and largest technical university in Germany and one of the major players is TU Berlin. Within its Center for Entrepreneurship, a successful incubation program for high-tech startups has been created.

1.1 BERLIN: STARTUP HUB NO. 1 IN EUROPE

Berlin, capital of Germany, is a growing creative and dynamic startup hub. Nowadays, the startup hype brings a vital entrepreneurial culture in Berlin.

Stock exchange listed companies open up accelerators, hubs, and co-working-spaces – all of them trying to find the next breakthrough innovation. Innovations and startups are promoted by the government since a decade. The city has four research universities and six applied science universities with around 165,000 students. Major research institutions, such as the Max-Planck-Society and Fraunhofer Society, are located in and around Berlin. The city also has two major technology parks, Campus Berlin Buch and Adlershof, the latter being one of the largest science and technology parks in the world. So there is a huge pool of talent available (Mrozewski et al., 2016). People from all over the world come to Berlin and start a business. The cost of living, renting offices and hiring talent is lower than elsewhere. The dynamics and creativity in the city may have their roots in political and historical upheavals of the 19th and 20th centuries. Virtually no other major city has reinvented itself as often (www.tu-berlin.de). Startups from academia play a major role within this economic system. They contribute to wealth creation, mostly in the service and information sector. These are quite crucial due to a lack of industrial jobs. Even less than 10 years ago, the promotion of startups at the universities was not really a serious issue. Nowadays, most of the universities and colleges in Berlin established appropriate facilities to foster entrepreneurship education and practical support (Mrozewski et al., 2011).

1.2 TECHNISCHE UNIVERSITAT BERLIN: WE HAVE THE BRAINS FOR THE FUTURE!

Technische Universität Berlin (TU Berlin) has been established in 1770. Nowadays, it sets a focus with its seven faculties on engineering, life sciences, and businesses as well as social and planning sciences. With more than 34,000 students, 350 professors, 2000 research scientists, 40 bachelor degrees, and 63 master degree programs, TU Berlin is one of the biggest technical universities in Germany, internationally known and rich in tradition. Well-known inventors and founders have their academic background at the TU Berlin: Konrad Zuse, who invented the first programmable computer system; Ernst Ruska, creating the first electron microscope and winning the Nobel Prize for physics in 1988 as well as Max Schaldach, building the first heart pacemaker. TU Berlin is the largest university with a technical focus in the capital region. It takes an outstanding role in the field of enterprise foundation and technology transfer in Berlin. For more than 25 years, it aims at bringing forward enterprise foundations from its own ranks, establishing the first high-tech incubator in Germany in 1983. TU Berlin aims to develop Science and Technology into useful solutions for society. Members of TU Berlin are committed to sustainable developments which both solve current challenges whilst not compromising or burdening future generations (www.tu-berlin.de). TU Berlin believes that research and education are inseparable and promotes the transfer of both knowledge and technology from inside the university out into practical applications. It supports the transfer of innovation through the creation of startups and businesses founded by students and scientific staff. About 1,200 founders are part of the alumni program – a remarkable number in the German area. There exist clear guidelines for the handling of intellectual property rights like patents and founder have the first choice for buying opportunities. TU Berlin has been awarded as "EXIST – Entrepreneurial University" as one of six universities out of 150 universities in Germany.

1.3 CENTER FOR ENTREPRENEURSHIP: ENTREPRENEURIAL EDUCATION, CULTURE, AND LOCATION CENTER

Strengthening the startup eco-system within TU Berlin is the main focus of the Center for Entrepreneurship (CfE). It offers the best training programs for intrapreneurs and entrepreneurs in Germany since 2011. It plays a major role in entrepreneurship research and practice throughout Europe. As a university startup incubator, it supports the transfer of technology-driven innovations into new businesses, thereby benefitting society as a whole.

With direct access to the substantial resources of the seven faculties, CfE develops startup ideas and startup teams whilst nurturing the startup mentality within a technological and knowledge-based environment. It offers a wide range of hands-on courses and classes about self-employment, startup days, speeches, and workshops as well as individual coaching and established tools to develop the entrepreneurial mindset and motivate entrepreneurial action. CfE's activities have a scientific background and we use state-of-the-art tools, essential for ensuring success. CfE combines expertise in research, education, and startup support (Mrozewski et al., 2016). Here, all stakeholders involved in entrepreneurship meet as part of a highly developed network, including members of the university with an interest in this topic, alumni founders, teachers, investors, business angels and many more. The service and programs are addressing students, graduates, and scientific staff, motivated in founding an enterprise.

The CfE is led by Prof. Dr. Jan Kratzer, Department of Entrepreneurship and Innovation Management and Dr. Florian Hoos, director of the StarTUp service, one division of TU Berlin's department for research support services. It is a formalized strategic collaboration between the theoretical and the practical side of entrepreneurship, bringing fruitful insights due to consistently exchanges (see Figure 1.1). It is part of the executive department and in direct contact with the president of the university. This structure is quite unique in Germany.

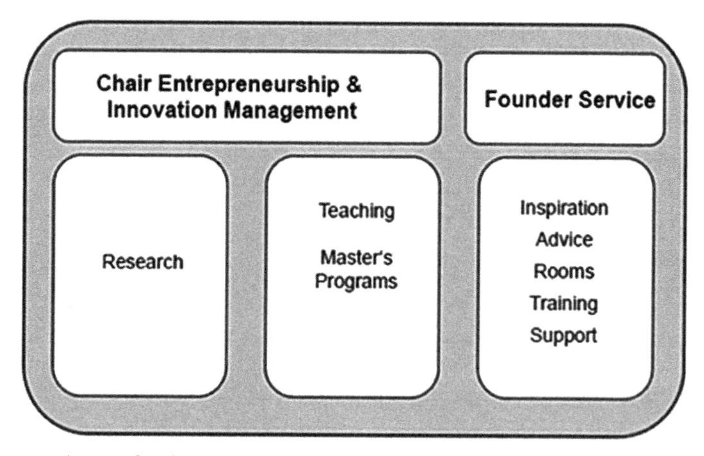

FIGURE 1.1 **(See color insert.)** Organizational structure of center for entrepreneurship.

1.3.1 INTERNATIONAL DUAL MASTER PROGRAM

The chair of Entrepreneurship and Innovation Management offers a two-year double master degree program at Technische Universität Berlin and Twente University or Warsaw School of Economics. The Master of Science (MSc) Innovation Management, Entrepreneurship, and Sustainability enables students to study at two locations and receive a master degree in Innovation Management and Entrepreneurship from TU Berlin as well as a master degree in Business Administration from Twente University or a master degree in Management from Warsaw School of Economics.

Students on the Innovation Management, Entrepreneurship, and Sustainability program will learn above all the fundamental principles of innovation management and entrepreneurship. This includes all issues on operative, tactical, and strategic levels relevant to founding new businesses and in R&D. Graduates of the program are able to tackle issues independently, assess research positions and findings competently and in detail and apply them to their own research. This is achieved by linking a range of topic areas from economics and social science.

Graduates of the Innovation Management and Entrepreneurship study program are employed in many areas of the economy and in public service. The program-wide range prepares them for a wide variety of professional fields. Some typical professional fields include: innovation management, research and development, teaching in the academic field, marketing, organization and company planning, human resources, environmental management, management consultancy, and corporate management. These professional fields also offer a range of starting points for setting up one's own company later, e.g., in management consultancy or through partnerships with other companies. The practice-related aspects of the program include – among others – gaining knowledge of structures, processes, legal provisions, and so forth. Teamwork is also increasingly promoted.

1.3.2 INCUBATION PROGRAM AND OTHER PRACTICAL SERVICES OFFERED

CfE provides services aimed at a successful market entry for innovative high-tech founders. With a 12-months incubation program, including an in-house business model design workshops following the business model canvas by Osterwalder & Pigneur (2010) and team building support. CfE helps potential founders applying for startup grants for perusing an innovative technology-oriented business idea from the state government as well as federal government. The pre-seed grants can be applied by founder teams with three team members and a professor as a mentor. CfE also provides facilities and office spaces with the startup incubator and prototype workshop facilities. Within the first 18 months, the office space is free of charge for the potential founders.

Due to the tradition of fostering technology transfer, TU Berlin has established an alumni network with more than 1,200 members related to startups. This is quite excellent in the German university landscape. About 100 of them are engaging in the startup support services at the university (e.g., as mentors, investors, lectures, experts in business modeling). Every year, a representative alumni network event is held at CfE and all stakeholders come together.

For every two years, a startup survey is conducted among the startups and companies related to TU Berlin. About 302 companies took part in 2016, and they make a total turnover of 2.6 billion Euro and employed around 18,400 persons in 2015. Therefore, the economic value of the venture from the university has an important impact on the region. About 80% of the companies are still located in the area of Berlin.

Furthermore, CfE has built up an own investors club, a fruitful network to venture capital companies and business angels, helping the forthcoming startups getting seed funded. Alumni are among the investors. Another evolving part of the support process is bringing young business creators together with big companies, industries to find pilot customers or manufacturers.

The startups supported by the CfE are active in the field of software, information and communication technology, technical consulting and research (including architecture and planning), and as well as non-technical consulting and research. About 66% of the startups are related to the service sector. This is quite important due to shrinking industrial companies in the region of Berlin. Most of the startups are profitable in the first year.

CfE provides with its incubator a protected area for the sprouting business ideas. There is no acceleration and profit orientation towards the startups. All services are offered for free and there is no legal form to share in the companies. The main focus is to bring useful innovations to society to improve economic and social prosperity.

For bridging the gap between basic research and applied innovation, TU Berlin created an incubation program in 2014. It is oriented by the programs on the existing incubators and accelerators like ‘Y Combinator’ and bundled all support offers for early-stage high-potential. Founders with really innovative high-tech business ideas can peruse a systematic process to bring the idea to market.

CfE helps potential founders applying for startup grants from the state government as well as federal government. Thanks to the national support programs of the ministry of economy and technology, the EXIST-scholarship and the research transfer, spin-offs from universities are systematically supported in financial terms. Within those programs founder teams, which in most cases consist of research assistants, have to come with a professor in the role of a mentor. Business ideas have to be innovative and should be based on the research activities of the university. The support is mainly focused on the development of a business plan including the preparation of the market entry (EXIST Business Startup Grant Euro 112,000) or on complex and risky research and development activities (EXIST – Transfer of Research, approx. EUR 600,000–1 million for R&D and market entry). The pre-seed grants can be applied by founder teams with three team members perusing an innovative technology-oriented business idea. Most of the founder teams passed through the process, have received the EXIST-Business Startup Grants from the government. Pre-seed funding is a requirement for incubation.

In the following, we describe the process of the support service. In the beginning, there are a few formal requirements to fulfill to be able to apply. These are:

- Innovative technology-driven business idea;
- The first prototype exists;
- Wanting to found in a team;
- Contact with the research department at TU Berlin;
- No formal founding with that idea;
- University degree.

The application process for the incubation program starts with an appointment with our startup-consultants, a conversation about program details and the fitting of the business idea. Then, the potential founders have to pitch their idea in front of an expert jury. When there are accepted for the program, they have 15–18 months working on the validation of the business idea (see Figure 1.2). In the beginning, they create a range of business models following the method of Alexander Osterwalder. Our in-house experts are helping them to find a sustainable business model, working with experts from industries. Our consultants support finding team members. Further, the founder team has to develop a minimal viable product and go through some feasibility checks concerning legal matters. At that point the role of prospective customers to get feedback from the market and finding pilot projects is essential. Later on, the teams get prepared for pitching for investors (e.g., pitch training, verification of documents, financial strategy). A real advantage of the CfE is a private investors club and the access to a European network of investors and venture capital companies. The investor's dinner is the network event with VCs from Europe. The investments range between 200,000 Euro and 3.5 million Euro. The investors club combines private investors, which invest up to 150,000 Euro to leverage bigger investments.

Further, first business contacts with relevant industry companies are matched individually for every team. Through the whole process, the founder has to present their results several times, getting feedback for further directions. For the whole process, the teams can use free office space and prototyping lab facilities. An additional service offer is practical seminars for learning soft skills and state-of-the-art methods like design thinking, scrum, business model canvas, and as well as know-how in law and taxes.

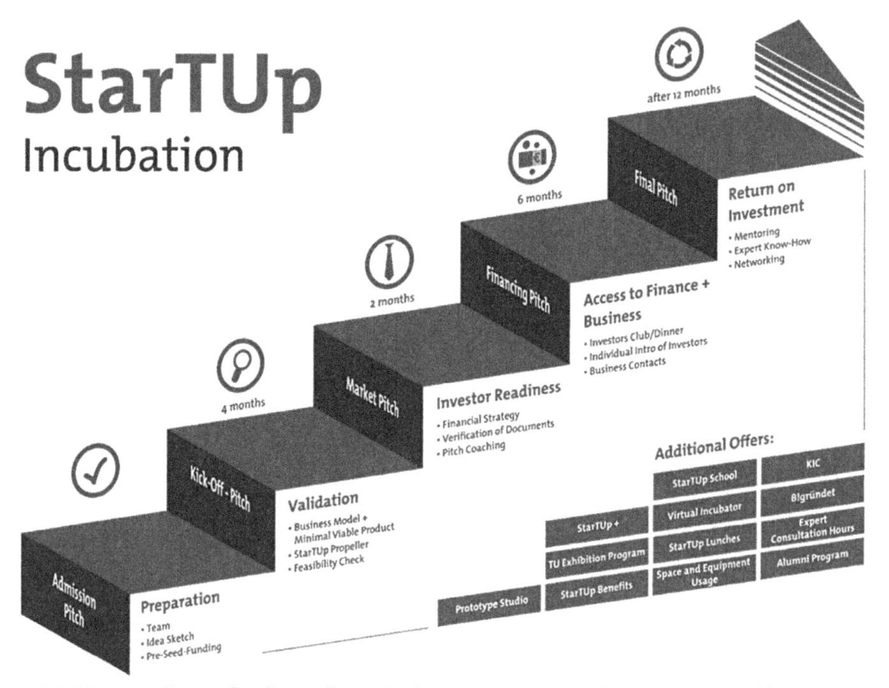

FIGURE 1.2 **(See color insert.)** Incubation program center for entrepreneurship.

About 40 startup teams are taking part per year, and about 25 companies are founding every year. Since 2007, 150 companies were supported by TU Berlin and 80% of them are still on the market. There are no cohorts or batches like in other accelerator or incubation programs.

1.4 CFE FACTS

- There is a high demand for the services and programs offered at CfE. The application-admission ratio is 7:1 students for the master program and 8:1 founder teams for the incubation program.
- Master of Science (MSc) Innovation Management, Entrepreneurship, and Sustainability is the most popular master program at TU

Berlin and the best entrepreneurship master programs in Germany (eduniversal-ranking).

- Top 2 universities with regard to raising pre-seed grants (EXIST-Business Startup Grant as well as EXIST Transfer of Research Grant) in Germany.
- High level of internationalization in teaching and applications come from almost 90 countries.
- For each Euro of public funding invested in the efforts at CfE, the supported startups raised 5 Euro pre-seed grants and seed funding.
- Eric Schmidt (Alphabet/Google) and Chris Cox (Facebook) inspired more than 1,000 students and 5,250 participants were welcomed at startup events in 2015.

1.5 STARTUP PORTRAITS

In the following, we describe shortly two of our supported startups in the area of business intelligence.

1.5.1 CEleraOne GMBH

CeleraOne, headquartered in Berlin Mitte, is one of the most interesting high-tech startups in Germany. A growing team of highly motivated, experienced software engineers develops efficient solutions to analyze and manage Big Data in real-time. They deliver complex IT projects on time. CeleraOne works closely with its clients to seamlessly integrate their products with the existing infrastructure. The team is exceptionally well connected and cooperates with renowned experts from various fields in information technology. The technology is built for performance, scalability, and the reduction of operating costs. They integrate all communication channels between the content provider and end-user, connecting with conventional enterprise solutions and free valuable business data from information silos. There are four key performance drivers for CeleraOne.

1.5.1.1 BUSINESS LOGIC IN REAL TIME

CeleraOne provides a complete analytics and automation platform optimized for synchronous real-time processing of people data, content, and device/machine signals. Through its unique Synchronous Insert-Compute-Respond

Operations (SICRO), CeleraOne provides instant results and control commands to connected users, devices, and machines within milliseconds, even under heavy loads. To achieve maximum flexibility without sacrificing performance, business logic is specified in the CeleraOne Programming Language (COPL), an easy-to-use, statically typed functional programming language that is JIT-compiled to native code. The use of COPL enables data to be processed on the fly while it is streaming through the CeleraOne system. Being a strictly functional language, COPL adds an extra speed advantage and a safety net for casual users who do not have a programming background. Instant computational results based on real-time business logic are at the core of advanced consumer and industrial internet applications.

1.5.1.2 IN-MEMORY STREAMING TECHNOLOGY

Time-critical applications of business logic require high-speed data access under heavy loads. CeleraOne stores all data in chronologically sorted streams of updates in main memory. Proprietary in-memory compression algorithms allow CeleraOne to keep months and even years worth of data in main memory. Storing incremental updates instead of squeezing data into rigid structures avoids the high maintenance overhead caused by schema migrations.

1.5.1.3 BUILT TO SCALE

Hardened under tough internet conditions CeleraOne's technology scales easily to the highest traffic volumes and data sizes. CeleraOne's platform is purpose-built for enduring scalable cluster operation and zero downtime. In most environments, CeleraOne systems do not need any scheduled downtime at all. For increased data safety, each data record is replicated to multiple in-memory nodes. Depending on the type of data and the dimensioning of the cluster different replication factors are used for even higher safety. Additionally, a redundant long-term persistence layer keeps all data on disk. Deployments of software updates are performed without system downtimes.

1.5.1.4 DATA INTEGRATION

CeleraOne includes standardized HTTP REST/JSON APIs for many business applications. For applications with the highest traffic volumes, the

ZeroMQ protocol can be used to stream data into the CeleraOne platform. Incoming data from different sources is validated and merged automatically. Tracking integration into web pages is a quick process using the standard CeleraOne JavaScript Client. A special variant for Smart TVs is also provided. For integration with mobile devices, CeleraOne provides native Software Development Kits (SDKs) for iOS and Android.

CeleraOne was founded by Dr. Falk-Florian Henrich (Mathematics, Informatics) in 2012 with its colleagues with background from Informatics and Accounting/Finance. They were supported by Prof. Dr. Peter Pepper, department of software technique and theoretical informatics. They were supported with office space and ESF-Stipendium (2011). For more information see www.celeraone.com.

1.5.2 MAPEGY GMBH

As a Berlin-based Big Data and Visual Analytics company, mapegy GmbH provides top insights from global innovation dynamics at your fingertips. Their unique algorithms retrieve global data that is innovation-related, i.e., from patents, science, press or social media. Superior indicators measure and track R&D. Top of the line visualizations let the customers explore the global innovation graph intuitively. Mapegy provides minimizing risks by accessing all relevant, global information, having facts and figures at fingertips before anyone else and superior, intuitive visualizations. Currently, they are a team of more than 15 smart entrepreneurs from all over the world – sharing a passion for digital intelligence and making their customers more successful. Mapegy has been founded by Dr. Ing. P. Walde (Informatics) and his colleagues with a background in Industrial Engineering and Informatics in 2012. Prof. Dr. Sahin Albayrak from TU Berlin, DAI-Labor, has been their mentor. They were supported with office space and the EXIST-Startup Business Grant as well the incubation program. For more information see www.mapegy.com.

1.6 CHALLENGES AND OUTLOOK

One of the challenge is the low rate of female participants in high-tech ventures. Less than 30% are female founders. We hope to increase the number to lift the potential. Another aim is to offer more international services, addressing global markets and investor networks to increase the success potential of the startups. Only 30% of the supported startups so far, are active in international markets. Another challenge is the get the students of the master program and lecture series Entrepreneurship into venturing. Only a few of them start a company by themselves. We are looking forward to the opening of our new 1000 m^2 co-working-space at the central campus in fall this year.

KEYWORDS

- **Berlin**
- **Center for Entrepreneurship**
- **incubation program**
- **startups from university**
- **Technische Universität Berlin**
- **technology transfer**

REFERENCES

EdUniversal Business School Ranking, www.eduniversal-ranking.com (accessed on 25 Oct 2018).

Mrozewski, F., & Von Matuschka, (2011). "The impact of technology-oriented university startups on regional development: The case of the technical university Berlin." In: Seliger, G., Khraisheh, M. M. K., & Jawahir, I. S., (eds.), *Advances in Sustainable Manufacturing – Proceedings of the 8th Global Conference on Sustainable Manufacturing* (pp. 63–68). Springer-Verlag Berlin, Germany.

Mrozewski, F., Von Matuschka, & Kratzer, F., (2016) "TU Berlin – an entrepreneurial university in an entrepreneurial city." In: Sven, H., & De Cleyn (eds.), *Gunter Festel: Business & Economics* (pp. 65–74). Academic spin-offs and technology transfer in Europe: Best practices and breakthrough models. Edward Elgar Publishing, Cheltenham, UK.

Osterwalder & Pigneur, (2010). *Business Model Generation: A Handbook for Visionaries*, Game Changers, and Challengers.

We've Got the Brains for the Future. www.pressestelle.tu-berlin.de/fileadmin/a70100710/Dokumentationen/Imagematerial/Broschueren/Imagebroschuere/TUB-Imagebroschuere_2013_engl.pdf.

CHAPTER 2

FIRSTBUILD: COMBINING THE INNOVATIVENESS OF A SMALL STARTUP WITH A LARGE CORPORATION'S STRENGTHS

SVERKER ALÄNGE[1] and ANNIKA STEIBER[2]

[1]*Department of Technology Management and Economics – STS, Chalmers University of Technology, Gothenburg, Sweden*

[2]*Menlo College, Atherton, California, USA*

2.1 INTRODUCTION

Observing that many good ideas taken all the way to prototypes ready for the industrialization phase never reach the market due to the inherent character of the large corporation. The research director of the household appliances division of General Electric realized that there was a need of changing the process of developing new innovative products that better satisfy the not yet defined needs of customers. Taking inspiration from a business startup culture, a unit called FirstBuild was created outside of GE, built on a logic of open innovation and with the intention to accomplish rapid learning through probe-and-learn with users. FirstBuild describes itself as an online and physical community dedicated to designing, engineering, building, and selling the next generation of major home appliances. It utilizes crowds for idea-generation, problem-solving, prototype development, and decision-making. The intention is, however, to create a way of working where the new independent unit also will benefit from having access to resources through deep ties with the mother organization and most importantly that the new unit, in turn, will be able to provide the mother organization with new but proven value offerings.

The purpose of this chapter is to describe the FirstBuild approach to solve the classical dilemma of simultaneously managing exploration and exploitation in large firms, and to relate the experiences and potential of this new approach to the literature on organizational ambidexterity and open innovation.

This empirical part of this chapter is a single case (Alänge & Steiber, 2018) based on an in-depth interview with a key person in the leadership group in combination with secondary data primarily accessed through the internet. As FirstBuild's business concept is to be extremely transparent, the platform can be accessed through the internet and different crowd-development and decision-making activities are possible to follow and comment upon online.

The chapter is structured as follows. First, a discussion of theory on ways of increasing innovativeness in incumbents while still keeping a focus on exploiting present strengths. Then, the findings from an empirical case study are presented. This is followed by a discussion of the empirical findings relating to the previously introduced theory, and the chapter concludes by addressing theoretical and practical contributions.

2.2 THEORETICAL FRAMEWORK

Large companies typically lose their innovativeness after sometime as new procedures and regulations aimed at strengthening key business processes in a growing organization simultaneously seem to stifle innovation. In times of discontinuous change, many previously successful large firms experience difficulties to innovate and even to survive (Christensen, 1997; Danneels, 2010).

In order to be successful and more cost efficient in mature high volume markets, process management, standard operating procedures and quality/lean approaches have shown to be of importance (Womack & Jones, 2003). But, in order to be innovative and develop something beyond the improvement of existing products and services, there is a need of pursuing strategies to cope with both the present business and with innovation (Tushman & O'Reilly, 1997).

Therefore, over the years a large number of approaches have been suggested and tested by large firms. For example, the localization of corporate research away from operations (Hounshell & Smith, 1988); the creation of separate corporate innovation units (von Zedtwitz & Gassmann, 2002); ambidexterity awareness focusing on both on organizational structure and the ability of top management to manage both the existing and the upcoming (Tushman & O'Reilly III, 1997); but also interaction with the outside world (Thompson, 1967); open innovation (Chesbrough, 2003); sponsored spin-offs/corporate spin-outs/corporate incubation (Rothwell & Zegveld, 1982; Weiblen & Chesbrough, 2015), corporate venturing (Hill & Birkinshaw, 2014), acquisitions (Stettner & Lavie, 2014), co-production/co-creation (Normann, 2001; Hughes, 2014); crowdsourcing from users and communities (Boudreau & Lakhani, 2013) and various forms of innovation hubs and startup programs outside-in or inside-out (Gawer & Cusumano, 2014; Weiblen & Chesbrough, 2015, Steiber & Alänge, 2016).

Numerous authors have identified the relative strengths and weaknesses of small firms vs. large firms when it comes to rapidly develop new ideas or to scale up operations to a large volume market, respectively (Weiblen & Chesbrough, 2015). Some have even suggested the obvious advantage of cooperation between creative startups and well-established market dominators (Mansfield, 1968; Rothwell & Dodgson, 1991). This has, however, shown not to be easy due to very different behavioral characteristics (Rothwell & Zegveld, 1982; Weiblen & Chesbrough, 2015) and many ventures have failed (Chesbrough, 2002; Hill & Birkinshaw, 2014).

In this section, theories on ambidexterity, boundary spanning, probe-and-learn, co-creation, and integration are introduced. Later, these will be used for the analysis in the discussion section.

2.2.1 ORGANIZATIONAL AMBIDEXTERITY

According to Teece et al. (1997), the winners in the global market have been firms that can demonstrate timely responsiveness and rapid and flexible product innovation, coupled with the management capability to effectively coordinate and redeploy internal and external competences. They refer to this source of competitive advantage as dynamic capabilities. For analytical purposes, Teece (2014) disaggregates dynamic capabilities into: 1) identification, development, co-development, and assessment of technological opportunities in relation to customer needs (sensing); 2) mobilization of resources to address needs and opportunities, and to capture value from doing so (seizing); and 3) continued renewal (transforming).

Dynamic capabilities are about being able to thrive when the context is changing and new opportunities need to be identified and acted upon. However, in order to survive and develop for the future, a company also needs to manage the present simultaneously. The ability to focus on both time perspectives at the same time has been called ambidexterity.

Companies constantly face the challenge of being excellent in satisfying today's customers while working on products and services for tomorrow—ones which could even be a direct threat to today's. March (1991) expressed this as being able to pursue both exploitation (of current business) and exploration (for the future). Tushman and O'Reilly III (1997: 14) argued that managers need to build ambidextrous organizations, which they described as "organizations that celebrate stability and incremental change as well as experimentation and discontinuous change simultaneously."

In line with Teece (2014), O'Reilly and Tushman (2013: 332) state that "... organizational ambidexterity (sequential, simultaneous, or contextual) is reflected in a complex set of decisions and routines that enable the organization to sense and seize new opportunities through the reallocation of organizational assets." Sequential ambidexterity means that organizations shift their structures in a sequential fashion over time as a response to changing needs of exploration or exploitation. In the face of rapid change, sequential ambidexterity might be ineffective and Tushman and O'Reilly (1997) argued for the need of simultaneous ambidexterity accomplished by structurally separated subunits for exploration and exploitation and for the need of an involved top management able to dynamically balance the two sides. Both sequential and simultaneous ambidexterity focuses on structural means to solve the exploration/exploitation dilemma. Gibson and Birkinshaw (2004) introduced a third category, contextual ambidexterity, which they defined as the behavioral capacity to simultaneously demonstrate alignment and adaptability across an entire business unit. Contextual ambidexterity is different from sequential and structural ambidexterity in its emphasis on individuals rather than organizational units making the adjustment between exploration and exploitation.

In the age of the Internet and cheap information processing, rates of change have accelerated and in many industries, the time has been reduced for the exploitation of existing products and services. This forces companies to increase their focus on exploration and development of both new offerings and new business models. IT and the Internet have also opened up new models of collaboration with entities and individuals outside the firm. This has led to the increased testing of new collaborative approaches to innovation, contributing to a greater variation in the new ideas that are being generated.

In accordance with a recent article by Benner and Tushman (2015), we see two major challenges for the entrepreneurial firm. The first is to deal with the "traditional" view on ambidexterity, of being able to juggle between exploitation and exploration, and the second is to address the issue of where innovation should be located— inside or outside of company borders. Benner and Tushman (2015: 11–12) argued that because of the dramatic reduction of communication and information processing costs and the increasing modularization of products and services triggered by digitalization and internet, the fundamental mechanisms and locus of innovation have shifted over the past decade. In their view, the intrusion of community or peer innovation shifts the locus of innovation from the firm to the community and it follows an associated open logic.

This view is also in line with O'Reilly and Tushman (2013) who comment that a promising domain for ambidexterity research is to move from the firm as a unit of analysis to the firm's larger ecosystem as the locus of innovation will increasingly shift to the community (von Hippel, 2005). They argue that future work on exploration and exploitation will have to include more awareness of the larger community, which will accentuate the need for research on capabilities in leading across boundaries as well as identity issues that span the firm/community borders.

2.2.2 PROBE-AND-LEARN PROCESSES

The demands on shorter development cycles have led to a trend towards putting products on the market rapidly in order to test concepts directly with users and potential customers to obtain immediate feedback (Brown & Eisenhardt, 1998). This kind of user testing is part of a probe-and-learn process (Lynn et al., 1996) in order to speed up the product development and create more customer value in much shorter time (Cole, 2001). Small startup firms typically use this kind of probe-and-learn process in order to develop products while simultaneously gathering data on an emerging customer base in order to convince potential investors in the next phase (Jolly, 1997; Alänge & Lundqvist, 2014).

Large firms that feel the pressure to raise their innovative capability try to emulate the work approach of small creative companies. In large firms, probe-and-learn processes have initially been introduced in some computer and internet companies (Brown & Eisenhardt, 1998). For example, Google tests product concepts on internal and external users in extremely fast test cycles (Steiber & Alänge, 2013). Other large firms realize the need of introducing more rapid learning processes but in practice, it has not been easy as it goes against the prevailing innovation culture in many organizations. For example, in companies with the ambition of keeping a stringent focus on doing right-first-time and limiting variation, probe-and-learn processes may be jeopardized (Steiber and Alänge, 2013b). Cole (2001) stressed that the generation of an error, which is part of 'probe-and-learn' processes for discontinuous innovation is a special challenge for the quality discipline "that has grown up viewing deviance and error as the enemy."

More recently, the ideas of Steve Blank and Eric Ries have popularized the lean startup approach focusing on a more systematic and validated learning to build a sustainable business. The fundamental thing in a lean startup is to build a (minimum viable) product, let customers test it and measure how they respond, and learn from this to either persevere along the current track or pivot (make a sharp turn) and try a new approach (Blank 2007; Ries, 2011). The lean startup methodology has been introduced not only for startups but also for public organizations and for large companies. One of these was General Electric's Energy Storage division that instead of a traditional product extension applied lean techniques searching for a business model and engaging in customer discovery in order to develop a new world-class battery manufacturing facility (Blank, 2013, p.72).

2.2.3 BOUNDARY SPANNING UNITS

Open innovation as a concept is relatively recent (Chesbrough, 2003) but the phenomena of exploration and exploitation over company borders are not new. In order to manage these innovation/commercialization processes, large corporations have over time developed various organizational units that serve a boundary spanning role (Thompson, 1967) in order to get access to ideas, technology, innovations, and competencies. Examples include corporate venture investment in small startups both to learn from and if successful considered for acquisition by the large firm (Weiblen & Chesbrough, 2015). Various mechanisms are being utilized, including licensing, joint ventures, acquisition, and localization of innovation units at science parks and in the vicinity of universities (Hounshell & Smith, 1988; Wallin & Lindholm Dahlstrand, 2007).

The strength of small high tech firms is in their ability to rapidly develop new ideas and test them on customers while their weakness is related to their limited ability to scale up operations for a large volume market. Large firms typically show the opposite areas of strength and weakness. That has led some authors to suggest different ways of cooperation between creative startups and well-established large firms where they play interactive and complementary roles (Mansfield, 1968; Rothwell & Dodgson, 1991). However, Hill & Birkinshaw (2014, p.1905) commented that corporate ventures suffer from a high failure rate and that in order to serve as a boundary-spanning unit for the large firm, the corporate venture itself need to be ambidextrous in order to provide linkages to various external actors and integrate their activities with those of technical core units.

Chesbrough (2002) commented that although large companies have long been aware of the value of investing in external startups "more often than not, though, they just can't seem to get it right." However, the acquisition of small firms is an important source of innovation, regardless of the company is acquired in order to develop a new business area, or being linked to an existing business area, or simply being acquired because of the competencies of its personnel (Steiber & Alänge, 2013a).

Chesbrough (2003) developed the *open innovation* concept based on empirical observations of inbound and outbound streams of technology to/from large firms. While some large firms mainly had kept its IP for defensive reasons, Chesbrough emphasized that alternative markets should be considered and that large firms need to balance the inbound and outbound flow. Relying on external technology and innovations also put new demands on large firms of developing an ability in managing and utilizing resources that the company does not fully control. This, in turn, leads towards an understanding of the need of cultivating ecosystems (Steiber & Alänge, 2013a) that can provide a massive contribution of innovative capability but also means sharing the wealth with others based on the assumption that the whole cake will become considerably bigger.

2.2.4 CO-CREATION

Co-creation is a concept that was initiated in the service context, as a service is in some way always recreated in the moment of production when the supplier meets the customer. Normann (2001) outlined and described co-creation processes where the user has an important role. The co-creation concept has increasingly been used for various innovation processes, including inputs from users, lead-users, and makers.

There has been increasing interest and several articles about specific ways to co-create during the idea phase (Hughes, 2014). Crowd-sourcing of ideas is increasingly practiced in industry, either with participation from within organizations, or by designated communities or open for anyone who likes to contribute with an idea. Sometimes co-creation includes Hackathons where idea providers also have the opportunity to go directly into a lab in order to develop a first simple demonstration prototype within a very short time period, e.g., 48 hours. There are basically two approaches to crowd-sourcing of ideas, either to present a clearly defined problem that needs a solution or to present challenge areas in search of new innovative approaches, i.e., a broader way of inviting to idea generation (Alänge & Steiber, 2017). Co-creation is, however, a concept with the much broader application. The initial use was linked to development with lead users, e.g., technically advanced users of scientific instruments (von Hippel, 1976). Later on, von Hippel (2005) and others broadened their focus to other advanced end-users, such as the surfers (von Hippel, 2005). In both cases, the innovation process is based on long-term relationships between the advanced lead users and the company launching the innovation on the market.

Another way of democratizing innovation is visible in the Maker Movement (Hatch, 2013). There are companies specializing in facilitating maker events including both building communities and establishing physical labs where makers for a very limited fee can access advanced metal and woodworking machinery in order to build functional prototypes. Mark Hatch, the co-founder of one of the pioneering makerspace organizations, TechShop, commented that: “when it costs $100,000 to fail, the money does not come out of our disposable income.... But now, investments, new product ideas, better mousetraps don’t cost $100,000 to develop – they cost $1,000” (Hatch, 2014, p. 37). On the book cover this is summarized as: “Average people pay a small fee for access to advanced tools ... All they have to bring is their creativity and some positive energy. Prototypes of new products that would have cost $100,000 in the past have been made ... for $1,000.” And Hatch (2014) “shares stories of how ordinary people have devised extraordinary products, giving rise to successful new business ventures.”

The idea of makerspace has also its parallels in the company world where some large companies have created labs or workspace for small firms in their vicinity. The idea has been to co-locate in order for the small firms to benefit from access to the larger firms competencies and for the larger firms to create relationships with small creative firms that at some point could provide essential inputs into large firms’ innovation processes, e.g.,

AstraZeneca BioVentureHub. For some large software firms, the rationale for co-location has been to stimulate developers to develop applications based on the large firm's software; in order to expand the market reach of the software and contribute to the ecosystem that the large firm is cultivating (Alänge & Steiber, 2017).

2.2.5 *INTEGRATION*

It has long been known that it can be difficult to bring new ideas and innovations developed externally into an organization and its business units and product development groups. This has been discussed as the Not-Invented-Here (NIH) syndrome (Katz & Allen, 1983). More recently, Lichtenthaler & Ernst (2006) argued that there is a need of recognizing not only the NIH syndrome but also other "syndromes" that can affect the knowledge acquisition, accumulation, and exploitation processes.

There are number of ways that have been used by companies for integrating new ideas/products (Alänge & Steiber, 2018):

1. the movement of people between the unit creating the innovation and the business unit – this is a way that has been successfully practiced in Japanese industry;
2. building personal relationships between units – selecting people with social competence and existing networks;
3. involvement of people from business units early on to create a feeling of ownership and to make the search for innovation more relevant for the business unit;
4. changing decision processes and incentive structures; and
5. leadership support on different levels to legitimize and put time and resources on integration.

2.3 DESCRIPTION OF FirstBuild

FirstBuild was created by General Electric in 2014 as an independent legal unit, in order to allow for product and business model development in an open innovation environment that is not restricted by the incumbent's culture and established a way of organizing. FirstBuild takes the inspiration from startup companies' nimbleness and ability to work in rapid development and learning processes. It has the freedom of running as a small startup that can run small initial batches of new product concepts, finding new ways to reach customers and obtaining immediate feedback from them. FirstBuild is a wholly owned subsidiary of GE Appliances (GEA) and its product area is home appliances. While GEA was acquired by the Hayer Group in 2016, FirstBuild remains a subsidiary and independent innovation unit for GEA.

2.3.1 FRUSTRATION AND INSPIRATION

The birth of FirstBuild was triggered by a frustration with the present way of doing product development in a combination with inspiration from the outside. It was a growing frustration inside GEA R&D that excellent product concepts that fulfilled identified customer needs still didn't reach the customers – if they were innovative and just a little bit outside of the existing core line of products and services. New entrants, competitors from Asia, on the other hand, became successful by entering the US market with offerings based on similar kind of solutions that remained in the laboratories at GEA. This was primarily attributed to a problem in managing financial risk – as GE, being a large incumbent at the US market, did not have good business models for managing financial risk for innovative new products. Second, in those cases a product reached the market, it took a very long time because of the decision-making process on whether the product would be scaled up or not. The reason is that the decision-making process was based on high volume thinking and brand-related demands on high reliability of all products. This contributed to a reluctance of putting products on the market that was not thoroughly tested and verified for large-scale production.

This frustration led to an interest in learning from startups who excel in the early phases of product and business model development and succeed to bring new products rapidly for testing in the marketplace. Studies of startup companies were combined with learning from workshops at GE where leading startup gurus (e.g., Steve Blank and Eric Ries) had been invited to present their ideas on "lean startups." Different units inside GE tried the lean startup methodology and reported some success Blank (2013). This provided an understanding of what was required for rapid learning processes with customers. But it also brought the realization that the road towards a more nimble way of developing new products and business models can be very challenging, if not impossible, inside a large corporation built around efficient processes for production and distribution on a mature market. The conclusion was that lean startup was interesting for GE but something additional was needed, a different operating model.

Inspiration came from meetings with representatives from the Maker-movement – that had shown that approaches to product and business model development that were very different from traditional ways, but still were perceived as viable. Local Motors, MakerBot, and TechShop all became important partners to cooperate with and learn from. Local Motors were viewed as an interesting partner as they are developing open source cars, i.e., a relatively complicated product. Local Motors uses an online platform to crowdsource vehicle-design and engineering ideas, which it prototypes in small open-access factories (https://localmotors.com/). By using in-house expertise to channel the wisdom of the crowd, the company has succeeded in much less time and to a fraction of the cost of launching open-source vehicles. This also led to a project for DARPA to design and prototype a military vehicle, which was accomplished in an extremely short time (4 months) and received positive press (e.g., SEMA, 2011) and it was even directly addressed by President Obama in a speech (Obama, 2011) – that brought GE's attention. The intention was to learn from Local Motors on how to do co-creating and a contract was signed between the two parties aiming at re-creating Local Motors' process. MakerBot is a 3-D-printer company with experience of developing an online forum (Thingiverse) where FirstBuild can post problems and projects that MakerBot users can work on. TechShop is a makerspace chain operating in several cities in the US which also provides an opportunity for FirstBuild to address practical problems/projects that the TechShop community can work on.

The initiation of FirstBuild was inspired by a Hackathon organized early in 2014 in cooperation with a local hacker group, LVL1, composed of artists, mechanics, IT, hobbyists, and GE engineers. The winner at this Hackathon "was an oven with a bar-code scanner capable of reading and perfectly executing cooking instructions encoded on packaged food.... To the executives at GE, Cprek's (the team leader) hack was a wake-up call. The idea for a bar-code-scanning oven had come up in internal ideas sessions before, and they knew it had great potential ... And yet the concept had never left the brainstorm stage at GE. ...That the executives were now staring at a working prototype of an idea they already liked – from an outside source – made them wonder how much innovation they were letting slide by" (Foster, 2014, p. 58).

Organized Hackathons have become part of the FirstBuild way of working, which can be illustrated by the yearly two-days event "Hack the Home"

Mega Hackathon, September 9–10, 2017 where 300 makers, engineers, programmers, tech enthusiasts, and designers are invited to come together in one location, collaborate intensively, rapidly prototype, build new ideas, and solve problems. They will have access to 3D printers, laser cutters, water jet, CNC mill, tool room lathe, sheet metal cutting and forming, welding, grinding, hand power tools, woodworking, reflow oven, and soldering. The winning submissions will share US$ 10,000 and hardware prizes (accessed 06/08/2017 from https://firstbuild.com/event/2017-hack-home/).

2.3.2 THE FirstBuild MISSION AND ORGANIZATION

The FirstBuild approach is to work as an independent brand in an extremely open interaction with external actors, without interfering with or from the GE Appliances (GEA) brand. FirstBuild has 23 employees in 2017; makers, brand ambassadors, engineers, and designers. The FirstBuild mission is according to its homepage (downloaded 06/08/2017) expressed as: Invent a new world of home appliances by creating a socially-engaged community of home enthusiasts, designers, engineers, and makers who will share ideas, try them out, and build real products to improve your life.

The business model is to create a viable product, manufacture it in a low volume, and test it on real customers willing to pay for the new product. These customers are the pioneers that are interested in trying a new innovative product well aware of what they are testing something that might need further quirking and that they can have an important role providing feedback to the product developers. FirstBuild's designated customers are the first 14% in Roger's (1983) diffusion model of adopters, where the first 5% give ideas (Innovators) and the next 9% (Early adopters) provide good feedback. The assumption is that based on feedback and market data from this 14% of pioneering customers, it would then be possible to "cross the chasm" (Moore, 1999) into a regular GEA business unit that would be responsible for scaling up and in order to reach the "early majority" in Roger's terminology.

Why would a regular business unit accept a new product that has been developed outside of its own control? As the product has been validated by the market when customers approve of it and are willing to pay a price, the end result for FirstBuild is that a small market has been established for the new product. FirstBuild can then go back to GEA with a lot more data on the new innovative product and its market. This is a major difference to what an internal R&D unit can offer, which typically is only an engineering prototype, without any customer feedback or actual market data. Actually, it could be more than just a product as it can also include experience and data from testing a new business model, e.g., sold and marketed differently, reaching a totally new target group, with a different cost structure, etc. All such 'features' of the new concept have been tested and FirstBuild can provide additional data for the GEA business unit.

One of the essential features of FirstBuild is its micro-factory, which at the FirstBuild homepage (downloaded 06/08/2017) is described as: A micro-factory is a place where your ideas can grow into real products. Through the use of advanced manufacturing techniques and rapid prototyping tools, products can be made on a very small scale up to the thousands. This enables products to quickly move from concept to creation to showroom floor. The FirstBuild Microfactory is divided into four sections: an interactive space for brainstorming and product demonstration, a lab for prototyping, a shop to fabricate components, and a building floor where products are assembled.

As the FirstBuild production volume is low, typically below 1,000–1,500 units, the financial risk is limited, which means that the product development time can be considerably shorter than if a large firm wants to introduce a new product. One example from FirstBuild is an idea that came up in September 2014 that was launched as a product on the market in January 2015, i.e., in 4 months, which is a large company setting would not even be possible to imagine. If a large company makes a mistake, the financial risk is large because of the sheer volume of production, and the result is a focus on processes, checks, and a lot of meetings. FirstBuild, on the other hand, is run like a startup company with very little processes, checkpoints or meetings – instead, it utilizes a combination of very experienced engineers inside and direct input from the outside. If a problem occurs, it is put on the website and someone somewhere outside can come up with an idea that FirstBuild can pick up and use. The facilities are also open so anybody can walk into the work area and come up with an idea or a device.

FirstBuild is thus extremely open and transparent and all projects are visible online (at firstbuild.com). Anyone can submit ideas and even build early prototypes with 3-D printers and the tools available in the FirstBuild workshop. Once posted, others can vote for or improve on the idea/design. If a design idea is considered promising by the FirstBuild community and the GEA leadership, it can become an official project where GEA engineers and designers can help develop a more advanced prototype that if feasible can go into low volume production by FirstBuild.

FirstBuild also expands its reach through its partners that have access to larger user/maker networks and online forums. While the consumers are the most important providing rapid feedback and input to the development process, startups, and entrepreneurs are almost as important as they are doing the same as FirstBuild – trying to validate value propositions on the market.

There are standard ways of paying for new ideas that individuals or companies contribute. If the idea is utilized by FirstBuild and put into production and reaches the market, then the idea provider gets 1% of net sales (e.g., of the value of a modification to an existing product or of the value of a totally new product) as royalties for 3 years for a non-exclusive license. This means that the idea provider can take the idea to any other place or application as FirstBuild never insists on patents or any kind of restrictions. The online site is not only for collecting ideas to solve problems, people that come are also customers that evaluate the ideas submitted and they provide product validation as the product development process proceeds.

Ownership rights are of no importance for FirstBuild because of the speed within the industry. The patent system's characteristics make it take too long time to get a patent, which opens up for new dynamics. For the actors in the home appliance business, speed is the most essential as new products' life cycles are short (3–4 years) in comparison to the time it takes to be granted a patent. In addition, the patent information is also a way of telling others what is on its way.

Earning money on small production by selling to a pioneering group of customers can also contribute to running R&D basically for free. However, if this will be possible to reach, needs still to be evaluated.

2.3.3 INTERACTION WITH GEA

Having the GEA still nearby is viewed as an advantage and something that can assist in crossing the chasm and avoiding the "valley of death." FirstBuild could be viewed as a "Fat Startup in the Rich Kid's World," i.e., that FirstBuild could afford to test approaches that could mean a threat of survival for a traditional startup. Although being an independent legal unit, it can still be difficult as the mother company keeps coming back and FirstBuild keeps fighting back as a rebel – "We spend 20% of our energy keeping the mothership away."

Still, the challenge is to scale-up within the mother company's business units – the lean startup work of developing the value proposition and establishing the business "is just 30% of the challenge." The FirstBuild unit allows GEA to test new innovative product ideas on end customers without jeopardizing its brand reputation, as the products are distributed and sold under a different brand. However, the subsequent integration of a promising and market-proven product into a GEA business unit is still a challenge related to the brand. The reason why is that brand is intimately linked to the existing supply chain and distribution channels, which means that the existing core brand name is a constraint because it chains the company to an existing business model with suppliers and customers. Large companies typically look for some form of validation in existing channels to the market, which in the case of GEAs home appliances could be the purchaser of a major customer, such as Home Depot.

FirstBuild still needs to develop its first product that it will bring back to the business units and it needs to develop the probe-and-learning process of finding ways of integrating new innovative products into existing business units. The operational model was easiest to change because it is mainly inside FirstBuild's control, while the business model development includes many aspects outside of direct control. In order to be successful, from the perspective of contributing to GEA's renewal, FirstBuild needs to be able to both keep/develop relationships and to contribute to changes internally in the mothership, which is a major challenge.

2.4 DISCUSSION

Large companies need to renew themselves in order to survive and many different approaches have over the years been suggested and tested in practice. This section will discuss how the design of FirstBuild and its first years experience compared to other approaches for renewal that can be found in the literature.

2.4.1 AMBIDEXTERITY BUILT ON DYNAMIC CAPABILITIES

FirstBuild is an example of simultaneous ambidexterity through structural separation by establishing an independent R&D unit that acts like a startup company (O'Reilly & Tushman, 2013). It was started as a separate unit under an active support by the GE CEO and the FirstBuild leadership continues to regularly report back to the top management of GEA. Thus, the establishment of FirstBuild follows the simultaneous/structural ambidexterity logic advocated by Tushman & O'Reilly III (1997), although the new unit formally is an independent company.

According to O'Reilly & Tushman (2013: 332) the appropriate lens through which to view ambidexterity is that of dynamic capabilities and they suggest "dynamic capabilities manifest in the decisions of senior managers, help an organization reallocate and reconfigure organizational skills and assets to permit the firm to both exploit existing competencies and to develop new ones." Ambidexterity viewed as simultaneously being able to pursue both exploration and exploitation requires both dynamic and static (ordinary/best practice) capabilities (Alänge, 1987; Tushman & O'Reilly III, 1997; Teece, 2014). Furthermore, dynamic capability defined as being able to sense, seize, and transform require a combination of technical (inventive) capability allowing for product/process innovation and entrepreneurial capability providing the basis for business model innovation (Alänge, 1987; Teece, 2014).

FirstBuild can be viewed as an R&D unit with entrepreneurial capability, which as a boundary spanner needs good relations both with the user/maker community and with its large firm initiator. Although FirstBuild has the good internal technical inventive capability, it is the crowdsourcing for ideas/prototypes/evaluations in combination with its backward links to development engineers in the mother organization that provides additional access to excellent dynamic technical capabilities. Thus, the FirstBuild approach enables the utilization of this 'dormant' engineering resource within the mother organization. Through its crowdsourcing network, FirstBuild also has a potential to source input for the development of new business models.

Large corporations in industrialized countries often have considerable technical (inventive) capability but due to path dependency and inertia lack the entrepreneurial capability to utilize this technology in new business models. O'Reilly & Tushman (2008) attribute this to senior managers' failure to capture the value from these technology resources. Sometimes the inventive technical capability is lacking as well, for example within the pharmaceutical industry where increasingly ideas and innovation come from small biotech firms that are being acquired by large firms. Thus, the problem is not only on the entrepreneurial side, although the lack of entrepreneurial capability has been singled out as one of the major barriers to innovation (Cowlrick, 2011).

2.4.2 PROBE-AND-LEARN PROCESSES

To act like a startup and test products on real customers is very hard to accomplish for large companies with an established brand that needs to be defended. Also the present distribution system and own business divisions expect to be part of something with substantial market potential (Alänge & Miconnet, 2001) – and typically make decisions based on a business case fabricated at the very start of a project – which is the opposite to a startup's probe-and-learn process, where both the technology/product/process and possible business models are developed gradually based on what is being learned over time (Jolly, 1997).

While a company like Google tests new product/service concepts on millions of users during a few days to get quantifiable data on user behavior (Steiber & Alänge, 2013), this is not standard practice in the home appliance industry. One reason is that the traditional way of marketing and distributing is geared towards mass markets, and not focused on the specific customers (innovators and early adopters) who could have an interest in trying something new and provides feedback to the producer (Alänge & Steiber, 2017).

FirstBuild, however, acts like a startup approaching the very early adopters (innovators) to collect useful feedback including an understanding of acceptable price levels (Ries, 2011). However, even if this early customer data is valid there is still a challenge to move a product from the startup environment into an existing GEA business unit – as such units normally rely on data from their traditional distribution channels.

2.4.3 BOUNDARY SPANNING

There are different boundaries in a corporation, both internal boundaries between different departments and the outer boundary to the external worlds. Boundary spanning individuals (technological gatekeepers) are essential for finding new knowledge outside and communicating this inside (Allen, 1977). Boundary spanning organizational solutions includes cross-departmental teams and technology transfer units (Alänge & Sjölander, 1986).

FirstBuild is boundary spanning on different levels. First, the transparent internet platform for new ideas for product and business model development opens up towards the external world, to smaller companies, users, and makers. Second, FirstBuild also has links backward to GE's business units. These are instead working according to a large-scale logic, which puts demands on other capabilities within FirstBuild. This means that FirstBuild needs to manage in two distinctly different worlds in an ambidextrous way (Hill & Birkinshaw, 2014). In the worlds of small-scale firms, users, and makers, FirstBuild is able to develop the product concept one step further than what would be the case within the regular R&D organization, including supplying customer data. However, from the perspective of its founder, GE, its value comes when it succeeds to integrate a new product concept into a regular business unit.

2.4.4 CO-CREATION FOR SHORTER LEAD-TIME

FirstBuild's way of organizing interaction with user/maker communities is built on a proven platform developed by Local Motors. This way of utilizing a verified "standardized" approach is a way of getting started more rapidly (Steiber & Alänge, 2015). By crowd-sourcing through Hackathons providing access to make space, ideas can immediately be developed into a first physical prototype. Also, the continued development inside FirstBuild is based on a crowd-sourcing principle by having a totally transparent process where the continued input to the development of an idea/product can come both through the internet and from persons walking through the physical FirstBuild plant. In addition, the crowd is also utilized for decision-making by being given the opportunity of voting for different proposals.

In principle, there is a possibility to either ask users/makers to solve a defined problem or to tell them about the challenges that the company/business unit is facing and ask them to provide ideas for solving generalized problems, without specifying what you are looking for (Boudreau & Lakhani, 2013). The former way can be useful for solving specific problems while the latter way allows for the unexpected to happen and for a broader input of ideas. The FirstBuild approach is typically the latter, looking for new innovative ways, but usually within certain limits given by the home appliances area. Whereas the original motivation to start FirstBuild was to provide the mother organization with innovations, some ideas/products generated, developed, manufactured, and sold to customers may face a more promising future outside of the large company's business units. Thus, they could be sold to other investors, e.g., to venture capital firms. Regardless of their final destination, through the probe-and-learn processes with real paying customers, FirstBuild is able to provide a base for decision-making including early customer data (cf. Jolly, 1997).

An interesting effect of the shortened product life cycles in home appliances (from 10 years towards 3–4 years), the role of IP has changed – speed becomes more essential than ownership of patents (Alänge & Steiber, 2017). This opens up for new strategies when it comes to interaction with idea providers as they can keep the IP and use it to develop other possible applications. This in turn means that an innovation unit like FirstBuild can be important in supporting the development of its own ecosystem, i.e., idea providers can learn from other applications based on their proprietary IP which in the long run further can benefit both the ecosystem and FirstBuild (Steiber & Alänge, 2013a: 589; Gawer & Cusumanu, 2014; Schmidt & Rösenberg, 2014: 78–79). In the case of FirstBuild, the rules for cooperation and IP ownership, as well as for remuneration are very clear upfront. This is something that small firms and developers ask for in order to enter a cooperation and it has become more common than large firms create clear rules upfront in order to limit subsequent conflicts. However, formalization can also become a barrier to collaboration when rules are introduced one-sided by a larger firm (Lawton Smith et al., 1991).

2.4.5 INTEGRATION OF THE INNOVATIONS IN THE MOTHER ORGANIZATION

It is clear that FirstBuild has found a way of generating ideas, developing product concepts and testing these on a limited customer base (Ries, 2011). From the perspective of a large firm looking for new products/innovations, the key issue is to make a decision whether the new product can fit in existing business units and if so, finding ways of integrating it in order to industrialize and scale it up to higher volumes. This integration is however not easy to accomplish in practice. There are many examples even of internal innovation units, working at arm's lengths distance in order to stay creative, having had difficulty transferring their new product concepts back to the business units. And the same NIH (not-invented-here) syndrome (Katz & Allen, 1983) situation exists for externally generated concepts.

For an independent but wholly owned unit like FirstBuild, it is important to continue cultivating relationships with the mother organization, including with senior managers and managers of the business units concerned. The employees need to be socially capable and well connected in order to fulfill the task of bringing their innovations back to the GEA business units. This indicates that FirstBuild as a boundary-spanning unit needs to be ambidextrous in itself (Hill & Birkinshaw, 2014). Developing an in-depth understanding of the challenges and potential needs of the business units and involving representatives from the business units early on in order to keep them informed and part of the process (Lichtenthaler & Ernst, 2006) are other integration mechanisms.

In a large company, new ideas can easily be killed by "knowledgeable" managers who know the present market. If an idea has a potential to be disruptive it might need early protection and the FirstBuild transparent idea generation approach through Hackathons with direct voting procedures offers a parallel decision-making structure that can help winners survive during early phases. This way of utilizing crowd-decision-making puts a demand on leadership that supports the process. In a Silicon Valley company this coaching leadership was expressed as: "trusting the judgment of competent people and supporting creative people even if I don't agree or fully understand what they see in it" (Steiber & Alänge, 2016). It is, however, not only the boundary spanning unit that needs to be ambidextrous and competent, a successful integration process also puts up demands on an absorptive capacity and suitable organization inside the large company (Cohen & Levinthal, 1990). FirstBuild considers it has a responsibility in not only delivering the new products/innovations but also in supporting the mother organization in developing such competence and structures.

2.5 CONCLUSION

The strength of small startups is nimble product/business model development processes with short lead-times while large companies typically excel in scaling up to large-scale manufacturing, marketing, and distribution. But is it possible to combine these strengths? The case of FirstBuild and the GE Household Appliances tells us that it can be possible to combine these strengths but also that it is not an easy task to accomplish.

FirstBuild functions like a startup in transparent interaction with users, makers, and customers, who can provide ideas and product concepts. FirstBuild also has access to engineering competence from GEA that can contribute both to the evaluation and further development of concepts. A key characteristic is the learning processes to collect hard facts from paying customers who also provide feedback for product development (Ries, 2010). The FirstBuild approach to interaction with users, makers, and customers is based on a model developed by Local Motors. According to this model, FirstBuild's capabilities, both inventive and entrepreneurial, are linked to a community based and transparent innovation process. This means that parts of the capabilities needed for the development of products and business models are externally localized. Thus, FirstBuild needs dynamic capabilities to be able to orchestrate a community of users/makers, small firms and venture capital.

In addition, FirstBuild aims to provide its mother organization GEA with radically new products. Here, the approach is less standardized and is still open to experimentation. There is always a risk that a new venture unit gets too much involved in exploration and searching for new innovations that are very different from earlier products. Then, it can be harder to integrate the new innovations to benefit business units in the mother organization. Hence, FirstBuild also needs to consider exploitation and thus, in order to meet each other, there is a need for ambidexterity in both the mother organization and in FirstBuild.

The FirstBuild approach emulates a small startup and by utilizing an approach and communication platform that to a certain degree are standardized, FirstBuild has succeeded to develop new products and tested them on a market with paying customers. From the perspective of the large corporation, FirstBuild is a boundary-spanning innovation unit that opens up for structural ambidexterity by both reaching out to a large community of makers and benefitting from its links to engineering resources in the mother organization. FirstBuild and its external communities have enough freedom to not getting caught in existing structures, in terms of distribution, customers, and brand – and has shown to have a capacity of generating innovative product designs. Some of these product innovations have also succeeded to get accepted by the mother organization's business units.

KEYWORDS

- **FirstBuild**
- **GE Appliances**
- **General Electric**
- **Local Motor**
- **Smith Corona**

REFERENCES

Alänge, S., & Lundqvist, M., (2014). *Sustainable Business Development: Frameworks for Idea Evaluation and Cases of Realized Ideas*. Gothenburg: Chalmers University Press.

Alänge, S., & Miconnet, P., (2001). *Nokia: An 'Old' Company in a 'New Economy.'* Paper presented at the 21st Strategic Management Society Conference, October 21–24 2001 in San Francisco.

Alänge, S., & Sjölander, S., (1986). *Strategisk Teknologitransfer Inom SAAB-Scania Combitech – del-Rapport 1*. Chalmers University of Technology and Institute for Management of Innovation and Technology (IMIT), Gothenburg.

Alänge, S., & Steiber, A., (2017). *Innovation Collaboration Between Small and Large Firms: Three Operational Models for Ambidexterity in Large Corporations*. Chalmers University of Technology, Gothenburg.

Alänge, S., & Steiber, A., (2018). *The Best of Both Worlds: Combining the Strengths of Small Startups With a Large Corporation's Advantages*. Chalmers University of Technology, Gothenburg.

Alänge, S., (1987). *Acquisition of Capabilities Through International Technology Transfer*. PhD diss., Chalmers University of Technology, Gothenburg.

Allen, T. J., (1977). *Managing the Flow of Technology: Technology Transfer and the Dissemination of Technological Information Within the R&D Organization*. Cambridge, MA: MIT Press.

Benner, M. J., & Tushman, M. L., (2015). Reflections on the 2013 decade award – "exploitation, exploration, and process management: The productivity dilemma revisited" 10 years later. *Academy of Management Review, 40*(4), 497–514.

Blank, S. G., (2007). *The Four Steps to Epiphany: Successful Strategies for Products That Win* (3rd edn.).

Blank, S., (2013). *Why the Lean Startup Changes Everything*. Harvard Business Review, 64–72.

Boudreau, K. J., & Lakhani, K. R., (2013). *Using the Crowd as an Innovation Partner.* Harvard Business Review, pp. 60–69.

Brown, S., & Eisenhardt, K. M., (1998). *Competing on the Edge: Strategy as Structured Chaos*. Boston, MA: Harvard Business School Press.

Chesbrough, H., (2002). *Making Sense of Corporate Venture Capital*. Harvard Business Review.

Chesbrough, H., (2003). *Open Innovation: The New Imperative for Creating and Profiting From Technology*. Boston, MA: Harvard Business School Press.

Christensen, C. M., (1997). *The Innovator's Dilemma: When New Technologies Cause Great Firms to Fail*. Boston, MA: Harvard Business School Press.

Cohen, W. M., & Levinthal, D. A., (1990). Absorptive capacity: A new perspective on learning and innovation. *Administrative Science Quarterly, 35*, 128–152.

Cole, R. E., (2001). From continuous improvement to continuous innovation. *QMJ, 8*(4), 7–21.

Cowlrick, I., (2011). *Drug Research and Development – Case Scenarios, Developmental Process, Risks, and Benefits*. PhD diss. Institute of Clinical Sciences, Sahlgrenska Academy, University of Gothenburg, Gothenburg.

Danneels, E., (2010). Trying to become a different type of company: Dynamic capability at Smith Corona. *Strategic Management Journal, 32*, 1–31.

Foster, T., (2015). Maker Inc.: How General Electric, Local Motors, and an army of DIY inventors are rebuilding American manufacturing. *Popular Science*, 56–72.

Gawer, A., & Cusumanu, M. A., (2014). Industry platforms and ecosystem innovation. *The Journal of Product Innovation Management, 31*(3), 417–433.

Gibson, C. B., & Birkinshaw, J., (2004). "The antecedents, consequences, and mediating role of organizational ambidexterity." *Academy of Management Journal, 47*(2), pp. 209–226.
Hatch, M., (2013). *The Maker Movement Manifesto: Rules for Innovation in the New World of Crafters*. Hackers, and Tinkerers. New York, NY: McGraw-Hill.
Hill, S. A., & Birkinshaw, J., (2014). Ambidexterity and survival in corporate venture units. *Journal of Management, 40*(7), 1899–1931.
Hounshell, D. A., & Kenly, S. Jr., J., (1988). *Science and Corporate Strategy: Du Pont R&D, 1902–1980*. Cambridge: Cambridge University Press.
Hughes, T., (2014). *Co-Creation: Moving Towards a Framework for Creating Innovation in the Triple Helix*. Prometheus: Critical studies in innovation, doi: 10.1080/08109028.2014.971613.
Jolly, V. K., (1997). *Commercializing New Technologies: Getting from Mind to Market*. Boston MA: Harvard Business School Press.
Katz, R., & Allen, T. J., (1983). Investigating the not invented here (NIH) syndrome: A look at the performance, tenure, and communication patterns of 50 R&D Project Groups. *R&D Management, 12*(1), 7–19.
Lawton-Smith, H., Dickson, K., & Lloyd, S., (1991). "There are two sides to every story": Innovation and collaboration within networks of large and small firms. *Research Policy, 20*(1991), 457–468.
Lichtenthaler, U., & Ernst, H., (2006). "Attitudes to externally organizing knowledge management tasks: A review, reconsideration, and extension of the NIH syndrome." *R&D Management, 36*(4), pp. 367–386.
Lynn, G., Marone, J., & Paulson, A., (1996). Marketing and discontinuous innovation: The probe and learn the process. *California Management Review, 38*(3), 8–37.
Mansfield, E., (1968). *The Economics of Technological Change*. New York, NY: W.W. Norton & Company.
March, J. G., (1991). Exploration and exploitation in organizational learning. *Organization Science, 2*, 71–87.
Moore, G. A., (1999). *Crossing the Chasm: Marketing and Selling High-Tech Products to Mainstream Customers* (Revised edition). New York, NY: HarperBusiness.
Normann, R., (2001). *Reframing Business: When the Map Changes the Landscape*. Chichester: John Wiley & Sons.
O'Reilly, C. A. III., & Tushman, M. L., (2008). Ambidexterity as a dynamic capability: Resolving the innovator's dilemma. *Research in Organizational Behavior, 28*, 185–206.
O'Reilly, C. A. III., & Tushman, M. L., (2013). Organizational ambidexterity: Past, present, and future. *The Academy of Management Perspectives, 27*(4), 324–338.
Obama, B. H. II., (2011) *President Obama Talks About Local Motors in a speech at Carnegie-Mellon*. https://www.youtube.com/watch?v=j_2zD-hs0aU (accessed 29 February 2016).
Ries, E., (2011). *The Lean Startup: How Constant Innovation Creates Radically Successful Businesses*. London: Portfolio Penguin.
Rogers, E. M., (1983). *Diffusion of Innovations* (3rd edn.). Free Press, New York, NY.
Rothwell, R., & Dodgson, M., (1991). External linkages and innovation in small and medium-sized enterprises. *R&D Management, 21*(2), 125.
Rothwell, R., & Zegveld, W., (1982). *Innovation and the Small and Medium Sized Firm*. London: Frances Pinter.
Schmidt, E., & Rösenberg, J., (2014). *How Google Works*. New York: Grand Central Publishing.

SEMA, (2011). President Obama recognizes Local Motors, DARPA, and American manufacturing. *SEMA eNews, 14*(26). https://www.sema.org/sema-enews/2011/26/president-obama-recognizes-local-motors-darpa-and-american-manufacturing-in-speec (accessed 29 February 2016).

Steiber, A., & Alänge, S., (2013a). A corporate system for continuous innovation: The case of Google Inc. *European Journal of Innovation Management, 16*(2), 243–264.

Steiber, A., & Alänge, S., (2013b). The formation and growth of Google Inc.: A firm-level triple helix perspective. *Social Science Information, 52*(4), 575–604.

Steiber, A., & Alänge, S., (2013c). Do TQM principles need to change?—Learning from a comparison to Google Inc. *Total Quality Management & Business Excellence, 24*(1–2), 48–61.

Steiber, A., & Alänge, S., (2015). Organizational innovation: Verifying a comprehensive model for catalyzing organizational development and change. *Triple Helix, 2*(14), 1–28.

Steiber, A., & Alänge, S., (2016). *The Silicon Valley Model: Management for Entrepreneurship.* Chem: Springer International.

Stettner, U., & Lavie, D., (2014). Ambidexterity under scrutiny: Exploration and exploitation via internal organization, alliances, and acquisitions. *Strategic Management Journal, 35*, 1903–1929.

Teece, D. J., (2014). The foundations of enterprise performance: Dynamic and ordinary capabilities in an (Economic) theory of firms. *The Academy of Management Perspectives, 28*(4), 328–352.

Teece, D. J., Pisano, G., & Shuen, A., (1997). Dynamic capabilities and strategic management. *Strategic Management Journal, 18*(7), 509–553.

Thompson, J. D., (1967). *Organizations in Action: Social Science Bases of Administrative Theory*. New York: McGraw-Hill.

Tushman, M. L., & O'Reilly III, C. A., (1997). *Winning Through Innovation: A Practical Guide to Leading Organizational Change and Renewal*. Boston, MA: Harvard Business School Press

von Hippel, E., (1976). The dominant role of users in the scientific instrument innovation process. *Research Policy, 5*, 212–239.

von Hippel, E., (2005). *Democratizing Innovation.* Cambridge, MA: The MIT Press.

Wallin, M.W., & Lindholm, D. Å., (2006). Sponsored spin-offs, industrial growth, and change. *Technovation, 26*, 611–620.

Weiblen, T., & Chesbrough, H. W., (2015). Engaging with startups to enhance corporate innovation. *California Management Review, 57*(2), 66–90.

Womack, J. P., & Jones, D. T., (2003), *Lean Thinking: Banning Waste and Create Wealth in Your Corporation.* Free Press, New York, NY.

Zedtwitz, M., & Von, Gassman, O., (2002). Market versus technology drive in R&D internationalization: Four different patterns of managing research and development. *Research Policy, 31*(2002), 569–588.

CHAPTER 3

THE END OF STRATEGIC LEADERSHIP: BE EXTELLIGENT OR EXTINCT

ROB VAN KRANENBURG

Founder of Council, Involved in NGI.eu (NGI Move and NGI Forward), Member of dyne.org and TRL, DeTao Master, Ghent, Belgium

3.1 DRIVERS

3.1.1 CHANGING NATURE OF POWER

Psychologists specialized in the behavior of larger groups of people try to explain the relative ease with which one is able to exert influence over masses by assuming "a causal force which bears on every member of an aggregate, and also for each individual there is a large number of idiosyncratic causes" (Stinchcombe, 1968: 67–68). He continues:

"Now let us suppose that the idiosyncratic forces that we do not understand are four times as large as the systematic forces that we do understand...As the size of the population increases from 1 to 100, the influence of the unknown individual idiosyncratic behavior decreases from four times as large as the known part to four tenths as large as the known part. As we go to an aggregate of a million, even if we understand only the systematic one-fifth individual behavior as assumed in the table, the part we do not understand of the aggregate behavior decreases to less than 1% (0.004)."

This shows how top-down power works and why scaling itself has become such an important indicator in such a system of 'success.' Imagine you want to start a project or 'do something' with your friends or neighbors, say, 5 people. This means that before you do anything – state a goal, negotiate deliverables, or even a first date on which to meet for a kick-off – that all five people relate to huge idiosyncrasies and generic forces that have to be aligned or overcome before you can even say 'Hello.' This shows how difficult it is to 'start something.' It also explains why you are always urged to get 'bigger' and why you need to 'grow.' It is only then and through the process of getting bigger itself that the management tools can operate, lying in waiting for you to 'discover them.' To be decisive, to make a difference, to set about a course for change, is in no need of 'growth,' nor in 'scaling.'

Understanding the nature of these social relations in the above terms show how difficult it is to script moments of systemic change, as hierarchical systems by the very fact that they are top down can concentrate on managing systematic forces relatively effortlessly. That which they cannot predict or control remains lone dissident, strange or abnormal voices, or 'sudden events.' With the internet, these idiosyncrasies have been able to organize and raise their weight in the ratio, and the Internet of Things will allow these even further, bringing the sensor network datasets individuals can handle to them on their devices. This acceleration of weak signals into clusters, organized networks and flukes cannot be managed anymore by formats that are informed by and that inform systematic forces as the nature of these forces has changed.

3.1.2 THE 'EXTRA' IN/OF TECHNÉ, DATAISM, AND MANAGING MOTIVATION

The efficiency paradigm fueled by a specific kind of 'life-form'[1] aka intelligence: the engineering maker/tinkerer optimizing what is at hand abstracts relationships between people, processes, and machines in behavioral models that can be broken down into trackable, traceable, and quantifiable properties.

This process has been instrumental to build the current techno landscape. "Enframing' was the process, according to Heidegger, that makes it 'normal' that for us that all things are 'standing reserve'; potentially ready to serve us as 'smart' objects. Technology then becomes more than a neutral tool, it becomes the very perspective of description itself. Even more, because it iterates and iterates in infrastructure it lasts longer than individual consciousness. Thus, an 'extra' is perceived – the accumulation of energy through mechanical and industrial means outside of pure human bodily capabilities, messing around in unpredictable ways with the notion of 'becoming' – innovation from pure human bodily tradition and experience. Hence, our current discussions and debate on the *unheimliches* of Artificial Intelligence, or whether robots will dominate us? It is important to realize that these words

[1] TIQQUN Introduction to Civil War [fragments]. The elementary human unity is not the body—the individual—but the form-of-life. http://www.softtargetsjournal.com/v21/tiqqun.php.

seem solid, but hide a full fluidity. What is 'us' is as malleable as 'robot.' And, 'is' is in real-time too.

3.1.3 DATAISM

The Shanghai city government released Honest Shanghai – you sign up using your national ID number – drawing on up to 3,000 items of information collected from nearly 100 government entities to determine an individual's public credit score. "A good score can lead to discounts, but a bad score can cause problems" (Rob Schmitz, 2017).

Jan Yoors writes how hard it was for him to be outside in the open for weeks on end. At times he longs for a door and to be able to lock it. The gypsies understand him, but for them, privacy is a state of mind: "...privacy was primary a courtesy extended and a restraint from the desire to pry or interfere in other people's lives...a mark of respect for them and of real compassion...." (Yoors, 1967).

We are a long way from this kind of thinking informing everyday practices. The default in building trust is no longer tribal in the sense of shared mental space. We should not lament this but propose building blocks for a praxis of everyday behavior that can help us build a shared framework that includes versions of the honesty app and the gypsy state of mind.

The result of dataism? The distrust of our analog senses, a steep price to pay. Not trusting your senses leaves you prone to a deep need to trust the datafied representations in apps, sensor readings, analytics over which you have no full control.

Its history runs as deep as the first classification attempts of Carl Linnaeus at taxonomy and Diderot and D'Alembert at breaking down knowledge in sequences of 'facts.'

3.1.4 MANAGING MOTIVATION

In *The Role of the Chief Transformation Officer,* Olivier Gorter, Richard Hudson, and Jesse Scott write in the *Recovery & Transformation Services,* November 2016, McKinsey & Company: "The individual charged with leading change must have multiple capabilities." McKinsey still thinks a "highly capable leader—the chief transformation officer" with compensation "linked to performance, with a significant bonus for over-delivery," should "behave like an extension of the CEO or even the board" as a "high-level orchestrator of a complex process that involves large numbers "of discrete initiatives." This approach is fundamentally flawed. The term strategic leadership itself has become irrelevant as in real-time there is no more viable place nor time (strategy-tactics). Just real-time.

3.2 TRAJECTORIES

The drivers above point to ontological change and fundamental disruption at all levels and domains of human production and consumption. This shift resembles that of 'fire,' 'book.' Still, most processes in business, government, and institutional frameworks act as if these changes brought about by these drivers can be contained within their current systems. A good example is the concept of the 'smart city.'

Around 2000, the IoT was termed by Kevin Ashton in such a way that unlike its predecessors – cybernetics (Wiener, Beer), ubicomp (Xerox Parc), pervasive computing (IBM) and Ambient Intelligence, AmI, (Philips) – hampered by the brand connotations for ecosystem uptake, that an ecology of systems could be built. What was clear quickly to those who understood its fundamentals were that its full disruptive nature would not leave one single human operation or institution intact that was build prior to the global protocol of pure flow (pass on the packet: TCP/IP). In a world of engineers and computer scientists in power data is a set of properties embedded in the very worldview that describes, determines, and evaluates the very nature of what is conceived to be 'data.' That data serves as a fundamental nurturing layer to derive information from.

This layer then gets added first into *industrial processes and concrete plants* doing predictive maintenance to cut costs, *household items* like lamps, fridges, televisions, and toothbrushes to make them interactive, upgradeable, and hubs or platforms for additional products and services (like the NEST and the videocam), *cars, planes, and trains* become connected as mobility patterns and transportation platforms and by extension the entire city is groomed in our minds eye as citizens to become a special kind of object itself, rather complex yes, but manageable. As such the smart city as a concept is the most complex product to come out of this efficiency paradigm that is the Internet of Things. In fact then, the smart city is a business model. Yet, in the discourse on the smart city, we can discern two broad categories: salespeople and romantics. The first group is unable to experience the very feeling as they rationalize their actions purely in business terms. The second group feels threatened to the core of their very existence in companionship with their surroundings and either attempts to flee to a no longer existing analogue territory, hack the superficial apparatus (cameras, botnet attacks, blockers) or resist with a logic that still thinks that notions of time and place matter in a real-time environment (they don't).

Based on the drivers above, there are three main trajectories for smart cities: gated communities, internet of neighborhoods and pragmatic cybernetics.

3.2.1 GATED COMMUNITIES

One of the main forms of building worldwide is gated communities. Because of the Internet of Things, there will be zones where citizens lease all services (housing, mobility, care, food, energy, schooling, policing...) for a fixed fee. As public spending is less and less, these zones will need more security, policing, and identity management control.[2] This concept is applied to cruise ships as well. At CES 2017 Carnival introduced the *Ocean Medallion3*. Capable of "whispering wirelessly to a web of onboard sensors" these wristbands anticipate guests' needs, keep track of roaming children, let the crew know where people are and where they might be heading. It also acts as a wallet, charging purchases to people's shipboard accounts, "with identities confirmed by photos that pop up on workers' screens and allow people to be greeted by name."[4] This trajectory can be described as gated military zoned communities and Mad Max in between as there are less and less public funding and less and less public agency by states that have already outsourced and privatized most services.

3.2.2 INTERNET OF NEIGHBORHOODS

The second trajectory is a growing grassroots involvement of Transition Town, local sharing initiatives, trends towards local ecologies of food and

[2] With the election of Donald Trump, this scenario becomes more likely in the US and in the rest of the world where this practice of building is growing rapidly. See https://www.amazon.com/Against-smart-city-here-Book-ebook/dp/B00FHQ5DBS

[3] Adult guests will be able to opt out of being discoverable by the devices, which are also designed to help people navigate from one place to another on board massive and sometimes disorienting ships. https://www.yahoo.com/tech/carnival-turning-cruise-ships-smart-cities-sea-142321985.html

[4] idem.

numerous digital social innovation campaigns to involve cheap solutions like LoRAWan, maturing Arduino and cheap microprocessors and sensors. This could lead to a local open source development of a balance between #IoT support of issues such as pollution, mobility, sharing of resources (energy, food, tools) and decision making (participatory budgeting).[5] The EU CAPS programs driven by Fabrizio Sestini of the European Commission have been instrumental in creating a vast and vibrant ecosystem of digital social innovation, DSI[6]. This trajectory can be described as the new iteration of a number of progressive movements since the 1960s shaping up as the 'sharing economy.' Currently, it is unclear if the millions of interpersonal positive interactions and transactions are not feeding only a relatively small number of actors who will bring more negative effects to the entire ecosystem in the longer run.

In my opinion, these two courses are inevitable and on their way. Agency is only possible within a course, not in diverting it. Policy focus should always be on the emerging actors, as only in the complexity of emergence the un-predictabilities are such that new courses of action appear 'possible'; acquire potentiality. The first trajectory is driven by fear. The second by naiveté. There is only one way to not only confront but resonate with the drivers that are real and present; to use them as building blocks to build a new political system based on radical transparency.

3.3 A PRAGMATIC CYBERNETICS

In 2015, during IoT Asia, I presented the results of the work of a small Council team.[7] The plan was to build a roof over Singapore making sure that all data sources from people, machines, and processes remained in a Singaporean Cloud. We proposed to make this feasible by using the passport – paper and a chip currently – as a device acting as an IoT controller to objects in the home, car, and city and as a gateway to ensure that end-users could only talk to a dedicated set of Singaporean platforms and Cloud services. Little did we know that Singapore, with a single line of government – was already thinking along these lines and deploying the fiber backbone

---

[5] This trajectory has never actualized before but in very brief moments like the Paris Commune, Autonomous Barcelona, Machno Ukraine. See http://www.situatedtechnologies.net/?q=node/108

[6] https://digitalsocial.eu

[7] Rob van Kranenburg, Usman Haque, Shuo-©-Yan Chou, Brian Yeung, Tom Collins, and Tania Grace Knuckey.

for such a plan. In March 2016, the Smart Nation initiative turns the island into a "living laboratory" – a kind of playground for testing smart solutions to urban issues. Dr. Vivian Balakrishnan, the country's minister for foreign affairs and minister-in-charge of the Smart Nation Initiative, claims: "There is much political angst about inequality and middle-class stagnation in developed economies," he said. "This has been accompanied by loud, populist, and ultimately futile arguments about yesterday's ideology and politics. ... In Singapore, we know that new technology trumps politics as usual."[8]

It looked like as given in Figure 3.1.

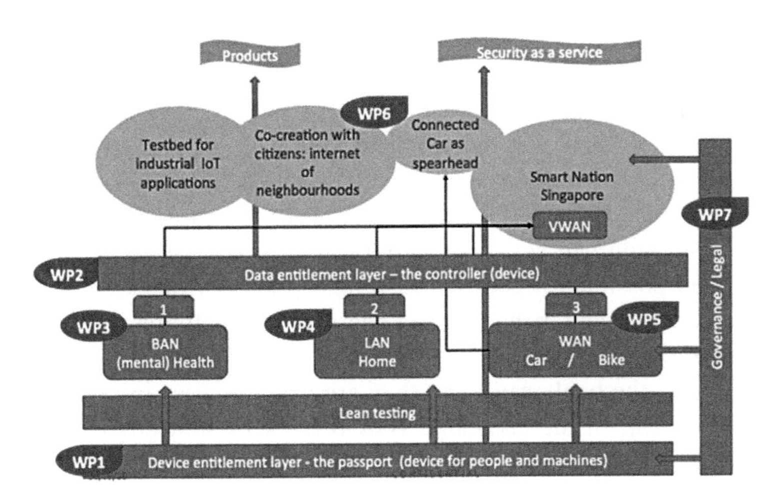

FIGURE 3.1 (See color insert.) Technology.

At the heart of this proposal lies a blockchain type of protocol[9] that identifies, authenticates, and authorizes devices to act on behalf of a service, a new kind of internet and a new kind of web. Security becomes a trusted service, unlike now as TCP/IP and HTML are fundamentally insecure protocols just saying 'pass on the packet' and 'pass on the link.' This system informs the

[8] Aaron Souppouris, @AaronIsSocial, Singapore is striving to be the world's first 'smart city.' Sensors, sensors everywhere. https://www.engadget.com/2016/11/03/singapore-smart-nation-smart-city/.

[9] "In a blockchain network the data is stored on many computers (the so-called 'miners') and every computer is directly in contact (a 'node') with all the other computers in the blockchain network. The information on all these computers is constantly aligned. Everybody owns and nobody controls the databases. The information is secured by sharing. In a number of cases, the blockchain software is even open-source software." http://www.thepaypers.com/expert-opinion/blockchain-for-dummies-a-quick-guide-into-the-ledger-technology/761925.

four key networks of #IoT: wearables and health in the BAN (Body Area Network), appliances, and services in the home (LAN, Local Area Network), telematics, mobility, and travel (WAN, Wide Area Network) and the smart city (Very Wide Area Network). All data remain with the citizens. The citizens entitle service providers to enrich, digitize, and personalize these data sets. In this system, citizens engage with their personal controllers – a dedicated smartphone doubling as a passport – through dedicated platforms in a single set of federated and potentially decentralized (using computers of locality) Clouds.

Apart from Singapore, such a system is being built in China and in many of the rising Asian tigers like Vietnam. Chinese #IoT top architects are laying the foundations for a pragmatic cybernetics in PR China. The new IoT reality organizes not around 'money' but around 'value.' It can first form value settlements, then find ways of sharing schemes. As such it can impact and disrupt four hundred years of capitalism, as it cuts out the middlemen then grow solid and rich and then start to protect their own positions forgetting their connecting, transitory nature. A faster and more precise model of value reformation can solve the current discrepancies in this equation. It will be then possible to receive a loan not on land or on assets but on the real motivation, talent, behavior, and demonstrated activities; very different but much more tuned to the time's parameters. Citizens are keeping RMB 140 trillion in the bank, saving as they are unsure where to invest. Putting #IoT at the heart of the economic system replaces subjectivity in pricing, brings real-time data to investments and thus restores confidence and trust. People who have money to invest enter into meaningful relationships with fellow citizens, who need money to make new infrastructure, services, and applications.

Franco Bruni, Economist, Vice President & Director, ISPI-Institute for International Policy Studies, highlights Rodrik's trilemma that democracy, national sovereignty, and hyper-globalization cannot go together: "So I maintain that any reform of the international economic system must face up to this trilemma. If we want more globalization, we must either give up some democracy or some national sovereignty. Pretending that we can have all three simultaneously leaves us in an unstable no man's land." Financial investments imply a correlation between a single party (Vietnam) or single layer (Singapore) of government and stability, claiming that democracy, as we understand it as a political multiple party process, is not productive at this moment in history.

In this scenario pragmatic cybernetics on state-level build capabilities for infrastructures and services to become smart. In this scenario, smart cities are the logical result of the political model.[10] "We want to make Shanghai a global city of excellence," says Shao Zhiqing, deputy director of Shanghai's Commission of Economy and Informatization, which oversees the Honest Shanghai App. Through this App., we hope our residents learn they'll be rewarded if they're honest. That will lead to a positive energy in society.[11]

In Europe, it is very difficult to find voices that understand that this is the only way to ride the transition. The Honest Shanghai App. is ridiculed. Citizens accept the semi-autonomous decision in their health, home, mobility, and smart city domain, but in politics, the narrative is fully drenched in emotional and socio-cultural, sometimes even religious arguments. The political arena is filled with populist voices and irrational stories. In the meantime, the gold of our times, data continues to flow out of Europe into Silicon Valley OTT databases; enriching them and making our national governments and EU Commission poorer by the hour. As they lose more and more agency, citizens – realizing the Emperor has no clothes – will start breaking away from the last public services and thus fuel scenario 1: Mad Max coming to you in under 10 years, unless we are able to build a European Cybernetics. Then, we need to co-create high level stimulating intellectual debate across Europe. An example is given graphically in Figure 3.2.

In this vision, Europe is a system of systems, a pragmatic cybernetics that eventually runs a 500 million European zone as a service. Compared to China we have the advantage of building a productive balance between extreme centralization on infrastructure, spectrum, smart contracts and hardware by EU industry and extreme decentralization on data fully open for direct local democracy and services and apps by EU SMEs; and personal data clouds as envisioned by the GDPR. Compared to Singapore we have the scale and the possibility to sell the system (running your territory-region as a service). Compared to the US we have relatively intact social systems and fully functional building blocks such as Estonian e-card, Horizon 2020, Digital Single Market focus. This scenario offers investors

---

[10] http://www.theinternetofthings.eu/aaron-souppouris-singapore-striving-be-worlds-first-smart-city'-sensors-sensors-everywhere.

[11] "A good score allows users to collect rewards like discounted airline tickets, and a bad score could one day lead to problems getting loans and getting seats on planes and trains." What's Your 'Public Credit Score'? The Shanghai Government Can Tell You, January 3, 20175:09 AM ET. http://www.npr.org/sections/parallels/2017/01/03/507983933/whats-your-public-credit-score-the-shanghai-government-can-tell-you.

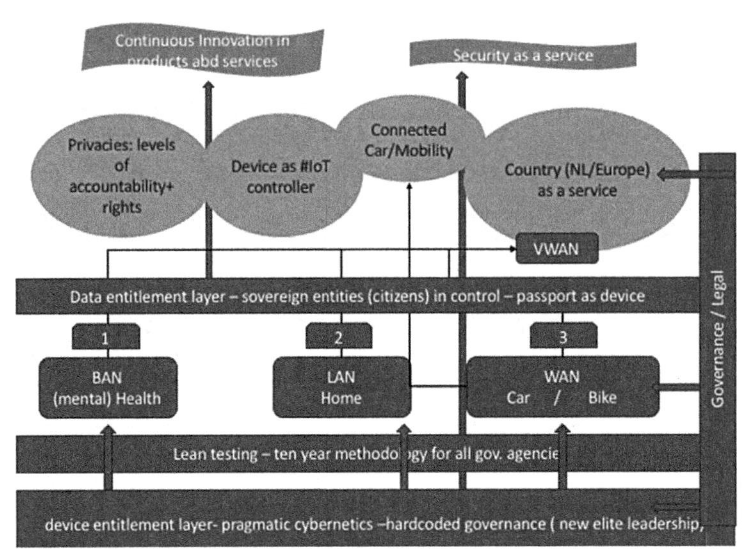

FIGURE 3.2 (See color insert.) European Cybernetics.

a real viable European project, strengthening public infrastructure and open business models on the data if they are on EU platforms, Cloud, and app store. We will disseminate, market, and crowdfund the scenarios, concepts, and stories broadly through the most penetrating channels of each of the target groups, national press agencies, and the appropriate EU media channels.

The example questions for co-creation are:

- What are the processes with which we can facilitate a transition to pragmatic cybernetics as a real-time decision-making system for a 500 million zone?
- What are the current mechanisms available to facilitate such a transition and what capabilities should be built in the short and mid-term to build a single European Cloud, a federated set of European IoT platforms and a dedicated set of protocols to hardcode Identity Management into a device (smartphone) that acts as a passport, controller of IoT devices, and payment infrastructure?
- What are the social and economic repercussions and instruments like basic income and how do these require border control of either territory or if the EU is run as a service, how do we define eligibility of

such a service (for example, the Estonian ecard is a service that does not require Estonian citizenship)?[12]

KEYWORDS

- **drivers**
- **gated communities**
- **pragmatic cybernetics**
- **trajectories**

REFERENCES

Jan Yoors (1967). *The Gypsies*. New York: Simon & Schuster.

Rob Schmitz (2017). What's Your 'Public Credit Score'? The Shanghai Government Can Tell You. http://spokanepublicradio.org/post/whats-your-public-credit-score-shanghai-government-can-tell-you.

Rob van Kranenburg, Usman Haque, Shuo-©-Yan Chou, Brian Yeung, Tom Collins, and Tania Grace Knuckey.

[12] The last section is part of an NGI, Next Generation Internet proposal by Mirko Presser, Jordu Artuff, Marta Arniani, Mannfred Aigner and Rob van Kranenburg.

CHAPTER 4

STRATEGIC DECISION TREES ON IMPACT COMPETITIVE MODELS

CYRUS F. NOURANI[1] and CODRINA LAUTH[2]

[1]*Acdmkrd-AI TU Berlin, Germany*

[2]*Copenhagen Business School, Digitalization Department and Grundfos Technology and Innovation, Denmark*

ABSTRACT

The new business analytics research area investigated in this chapter is concerned with how to combine novel *impact-driven* predictive modeling techniques using a blend of strategic decision systems. Different applications for advances in business analytics, enterprise modeling, cognitive social media, are developed with a more intuitive and self-organizing framework for designing *impact-driven* business interfaces that are more plan-goal-oriented business infrastructures. Dynamic business context requires that we look at more flexible and *impact-driven* plan-goal-decision-tree satisfiability rates combined with strategic and competitive business models. These new impact-driven predictive models can be flexibly integrated into new and complex information systems as well as in highly heterogeneous digital infrastructures for designing new impact-driven competitive business models on the fly. Additionally, these advanced business analytics techniques can serve as intuitive business process modeling techniques enhanced with the strategic and *context-driven* design of the predictive analytics tasks.

4.1 INTRODUCTION

As digital infrastructures and digital business services are becoming more and more complex and the pace of change for digital innovations is dramatically increasing, it becomes highly inefficient to use only traditional data-driven predictive modeling to enhance complex and strategic business modeling. The novel applied research areas explored in this chapter range from plan goal decision tree satisfiability with competitive business models to predictive analytics models that accomplish goals on a 3-tier business systems design models. Original work on attention spanning trees are applied here specifically to focus on plan goal models that can be processed on a vector state machine coupled with a database pre-processor datamining interfaces (Modeling, objectives, and planning issues are examined to present precise decision strategies. Competitive decision tree models are applied to agent learning. Enterprise systems stage sequence communications with business objects and basic content management with multi-tier interfaces are being explored, focusing on attention spanning (Nourani, Lauth & Pedersen, 2015) with specific examples developed in Lauth, Nourani, Pederson & Bloome, (2013) and Lauth (2013).

The field of automated learning and discovery has obvious financial and organizational memory applications. There are basic applications to data discovery and model discovery. A competitive business modeling technique, based on the first author's planning techniques are stated in brief. Systemic decisions are based on common organizational goals, and as such business planning and resource assignments should strive to satisfy such goals. Heuristics on predictive analytics are examined with brief applications to decision trees. Russel-Norvig heuristic, an important criteria instantiations for the techniques, is presented in brief.

Software agents are specific computing agents designed by agent languages to specify how tasks can be relegated to agents to characterize a software functionality. There is agent computing, cyberspace computing, intelligent multimedia and heterogeneous computing. Plans and goals (see the preceding chapters) are applied to business planning (Nourani, 1998a,b) (Figure 4.1).

The basic multi-tiered designs are based on the following layers. The presentation layer contains components dealing with user interfaces and user interaction. Example of a visual JAVA standalone application. A business logic layer contains components that work together to solve business problems. The data layer is caused by the business logic layer to persist state permanently. More and more enterprises recognize that in the electronically archived databases a there is a potential for knowledge that could be processed up to now only insufficiently. An example application for competitive models appears in the transactional business models. Alternate models can be designed based on where assets, resources, and responsibility are assigned; how control and coordination are distributed; and where the plan goals are set. A transactional

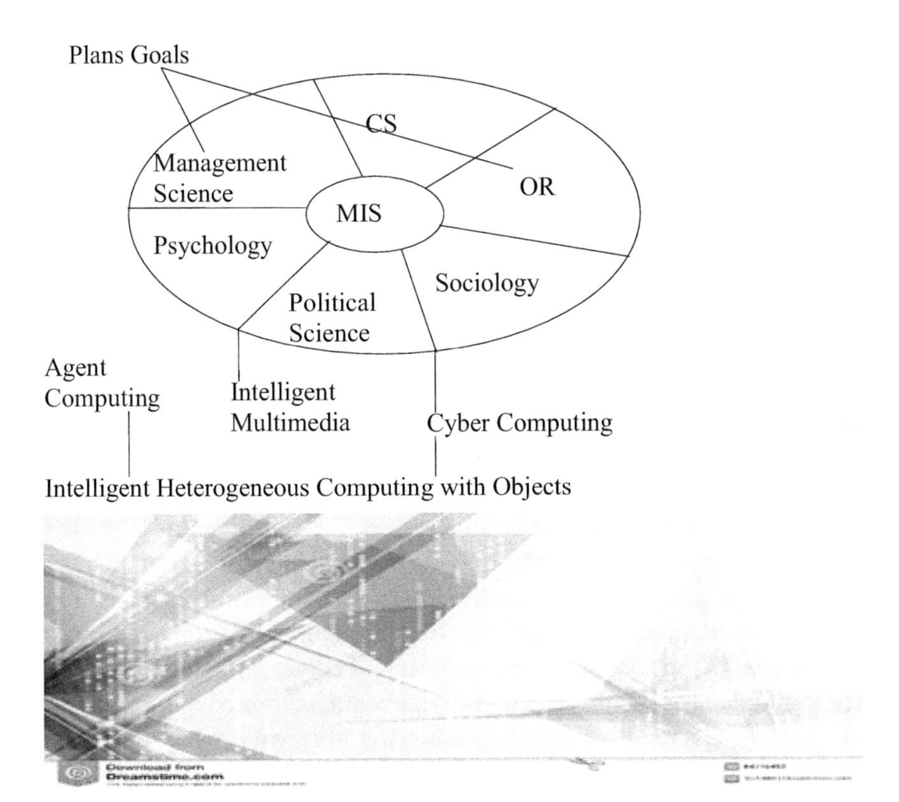

FIGURE 4.1 (See color insert.) New MIS areas.

international business model might comprise a coordinated federation with many assets, and resources. The overseas operations are considered a subsidiary to a domestic central corporation. However, the decisions and responsibilities are decentralized. Administrative formal management planning and control systems are how headquarters-subdivision controls are managed.

The corporations need to specify how to conduct business before they implement an ERP system. The best way to do that is by defining how decisions are made in terms of planning. Applying ERP to the planning process, we might develop tactical planning models that plan critical resources up to sales and delivery. Planning and tasking require the definition of their respective policies and processes; and the analyses of supply chain parameters. Dynamic thinking, that the business world is not static, is a fundamental premise to ERP. Operational thinking is applied when planning for decisions based on how things really work and interact. Closed-loop thinking can be applied to control decision trees on ERP when feedback is necessary to carry on new plans or to set new business goals. Experience Management (EM) at times can assist in obtaining closed loop control or be applied to determine the scope of the project to fit within the resources available (including budget) and the time required and to assign responsibility.

ERP has to select the operational planning teams that will manage the process. Amongst the tasks are: feasible production schedules, mapping the operational planning process to develop scheduling models for production and plan inventories. Develop a plan for building a data warehouse, KM, and interfaces to legacy systems and ERP. ERP involves implementation teams consisting of business representatives, functional representatives, IT personnel and system knowledgeable people.

The section outlines are as follows: Section 4.1 is about the introduction. Section 4.2 presents the basics on competitive goals and models. Agent AND/OR trees are applied as primitives on decision trees to be satisfied by competitive models. Planning with predictive models and goals are presented with stock forecasting examples from the first author's newer decade's publications. Section 4.3 briefs on goals, plans, and realizations with data-based and knowledge-based. In that context a function key interface to the database is presented with applications to model discovery and data mining. Competitive model goal satisfiability with model diagrams is briefed with examples. Section 4.4 presents the applications to decision trees and practical systems design with splitting agent decisions trees. Cognitive spanning with decision trees and state vector machine computations applications are presented. Section 4.5 presents spanning applications with multitier business interfaces. Social media applications with Gatesense example from the accompanying author's is briefed. The chapter concludes with heuristics for competitive models and goals comparing with Russell and Norvig (2002) for decision tree accomplishment from the first author's newer game decision tree bases from Nourani & Schulte (2012–2014).

4.2 COMPETITIVE MODELS AND GOALS

Planning is based on goal satisfaction at models. Multiagent planning is modeled as a competitive learning problem where the agents compete on game trees as candidates to satisfy goals hence realizing specific models where the plan goals are satisfied; for example, Muller and Pischel (1994) and Bazier et al. (1997). When a specific agent group "wins" to satisfy a goal the group has presented a model to the specific goal, presumably consistent with an intended world model. For example, if there is a goal to put a spacecraft at a specific planet's orbit, there might be competing agents with alternate micro-plans to accomplish the goal.

While the galaxy model is the same, the specific virtual worlds where a plan is carried out to accomplish a real goal at the galaxy via agents are not. Therefore, Plan goal selections and objectives are facilitated with competitive agent learning. The intelligent languages (Nourani 1996, 1998) are ways to encode plans with agents and compare models on goal satisfaction to examine and predict via model diagrams why one plan is better than another, or how it could fail. Games play an important role as a basis to economic theories. Here, the import is brought forth onto decision tree planning. Newer tree computing techniques are applied to present precise strategies and prove theorems on multiplayer games. Game tree degree with respect to models is defined and applied to prove soundness and completeness.

4.2.1 AGENT AND/OR TREES AND IMPACT GOALS

AND/OR trees Nilsson (1969) are game trees defined to solve a game from a player's standpoint. Formally, a node problem is said to be solved if one of the following conditions hold.

1. The node is the set of terminal nodes (primitive problem: the node has no successor).
2. The node has AND nodes as successors and the successors are solved.
3. The node has OR nodes as successors and any one of the successors is solved.

A solution to the original problem is given by the subgraph of AND/OR graph sufficient to show that the node is solved.

A program which can play a theoretically perfect game would have tasks like searching and AND/OR tree for a solution to a one-person problem to a two-person game. An intelligent AND/OR tree is where the tree branches are intelligent trees. The branches compute a Boolean function via agents. The Boolean function is what might satisfy a goal formula on the tree. An intelligent AND/OR tree is solved if f, the corresponding Boolean functions solve the AND/OR trees named by intelligent functions on the trees. Thus, node m might be f(a1,a2,a3) & g(b1,b2), where *f* and *g* are Boolean functions of three and two variables, respectively, and ai's and bi's are Boolean-valued agents satisfying goal formulas for *f* and g.

The chess game trees can be defined by agent augmenting AND/OR trees (Nilsson, 1969). To obtain a game tree the cross-board-coboard agent computation is depicted on a tree. Whereas the state-space trees for each agent are determined by the computation sequence on its side of the board-coboard. Thus, a tree node *m* might be *f*(a1,a2,a3) & *g*(b1,b2), where *f* and *g* are Boolean functions of three and two variables, respectively, and ai's and bi's are Boolean-valued agents satisfying goal formulas for *f* and *g* (Figure 4.2).

```
g is on OR agent
      /|\
       |
   b1 | b2 f
     /_|_\
      /|\
f is an AND agent
    a1 a2 a3
```

FIGURE 4.2 Agent AND/OR decision trees.

We further define impact subgoals on decision trees to direct incremental decisions.

Definition: A plan subgoals on agent decision trees is an impact goal if f.*g* entails an essential goal for a plan.

The first author has considerable development on predictive modeling on game trees, e.g., (Nourani, 2000–2015) publications. Amongst the developments are a predictive explanation based reasoning and generalizations. That area applied the following since AI past decade: An explanation problem can be stated formally as: given facts Φ consistent formulas, known to be true, Defaults Δ: possible hypotheses, that we accept as part of an explanation and observations G: which are to be explained, An observation *g* in *G* is explainable, if there exit ground hypotheses ΩΔ, such that:

1. Φ, ∪Ω |= g, where |= indicates "logically implies," and
2. Φ, ∪Ω is consistent.

From the above, we can state the following: considering that a plan is a sequence on a decision tree to realize a goal G:

Proposition: Impact goals are explanations for a plan sequence of decision trees.

The proof explanations for this proposition are based on the first author's publications over a decade that are shown in part in a few chapters in this volume on planning and decision tree games.

4.2.2 PREDICTIVE MODELS AND BUSINESS PROCESS LANGUAGES

Predictive modeling is an artificial intelligence technique defined since the first author's model-theoretic planning project. It is a cumulative nonmonotonic approximation attained with completing model diagrams on what might be true in a model or knowledge base (Nourani, 1997). Prediction involves constructing hypotheses, where each hypothesis is a set of atomic literals; such that when some particular theory T is augmented with the hypothesis, it entails the set of goal literals G. The hypotheses must be a subset of a set of ground atomic predictable. The logical theory augmented with the hypothesis must be proved consistent with the model diagram. Prediction is minimal when the hypothesis sets are the minimal such sets.

4.2.3 THE STOCK TRADERS INTERFACE PREDICTIVE COMPUTING MODEL

The basis for forecasting is put forth at preliminary stages in the author's publications since 1998.

Schemas allow brief descriptions on object surface properties with which high-level inference and reasoning with incomplete knowledge can be carried out applying facts and the defined relationships amongst objects.

A scheme might be Intelligent Forecasting {IS – A Stock Forecasting Technique} Portfolios Stock, bonds, corporate assets Member Management Science Techniques.

We consider models as worlds at which the alleged theorems and truths are valid for the world. Models uphold to a deductive closure of the axioms modeled and some rules of inference, depending on the theory. By the definition of a diagram, they are a set of atomic and negated atomic sentences, and can thus be considered as a basis for a model, provided we can by algebraic extension, define the truth value of arbitrary formulas instantiated with arbitrary terms. A theory for the world reasoned and model is based on axioms and deduction rules, for example, a first-order logical theory. The above scheme with the basic logical rules, can form a basic theory T to reason about a stock forecasting technique. To do predictive analysis we add hypotheses; for example, from Nourani (2015), we have the atomic literals to complete the task.

p1. Asset (Stocks)
p2. Stock (x) => Asset (x) p3.
S&P100 (x) => Stock (x)

The predictive diagram for T is constructed starting with p1 = True, p2(f) = true for *f* ranging over stock symbols, p3 is true for all $x = f$, where *f* is a stock symbols in S&P100.

4.3 GOALS, PLANS, AND KNOWLEDGE BASES

Practical systems are designed by modeling with information, rules, goals, strategies, and information onto the data and knowledge bases, where masses of data and their relationships and representations are stored respectively. Forward chaining is a goal satisfaction technique where inference rules are activated by data patterns, to sequentially get to a goal by applying the inference rules. The current pertinent rules are available at an 'agenda' store. The rules carried out will modify the database. Backward chaining is an alternative based on an opportunistic response to changing information. It starts with the goal and looks for available premises that might be satisfied to have gotten there. Goals are objects for which there is automatic goal generation of missing data at the goal by recursion backward chaining on the missing objects as sub-goals. Data unavailability implies a search for new goal discovery.

Goal-Directed Planning is carried out while planning with diagrams. That part involves free of the plan that Skolemized trees is carried along with the proof tree for a plan goal. If the free proof tree is constructed then the plan has an initial model in which the goals are satisfied. A basis to model discovery and prediction planning is presented in Nourani (2002) and is briefed here. The new AI agent computing business bases defined during the last several years can be applied to present precise decision strategies on multiplayer games with only perfect information between agent pairs. The game trees are applied to improve models.

The computing model is based on a novel competitive learning with agent multiplayer game tree planning. Specific agents are assigned to transform the models to reach goal plans where goals are satisfied based on competitive game tree learning. The planning applications include OR- ERP and EM as goal satisfiability and micro-managing ERP with means-end analysis on EP. Minimal prediction is an artificial intelligence technique defined since the author's model-theoretic planning project. It is a cumulative nonmonotonic approximation attained with completing model diagrams on what might be true in a model or knowledge base.

4.3.1 DECISION-THEORETIC PLANNING

A novel basis to decision-theoretic planning with competitive models was presented in Nourani (2005) and Nourani & Schulte (2013) with classical and non-classical planning techniques (see, for example, Hedeler et al., 1990; Wilkins, 1984) from artificial intelligence with games and decision trees providing an agent expressive planning model. We use a broad definition of decision-theoretic planning that includes planning techniques that deal with all types of uncertainty and plan evaluation. Planning with predictive model diagrams represented with keyed KR to knowledge bases is presented. Techniques for representing uncertainty, plan generation, plan evaluation, plan improvement, and are accommodated with agents, predictive diagrams, and competitive model learning. Modeling with effector and sensor uncertainty, incomplete knowledge of the current state, and how the world operates is treated with agents and competitive models.

Bounds on game trees were developed based on the first author's preceding publications on game trees generalizations on VMK to on (Nourani and Schulte, 2013). Partial deductions in this approach correspond to proof trees that have free Skolemized trees in their representation. While doing proofs with free Skolemized trees we are facing proofs of the form p(g(.)) proves p(f(g(.)) and generalizations to p(f(x)) proves for all x, p(f(x)). Thus, the free proofs are in some sense an abstract counterpart of the SLD. Let us see what predictive diagrams do for knowledge discovery knowledge management. Diagrams allow us to model-theoretically characterize incomplete KR. To key into the incomplete knowledge base.

The Figure 4.3 depicts selector functions Fi from an abstract view grid interfaced via an inference engine to a knowledge base and in turn onto a database.

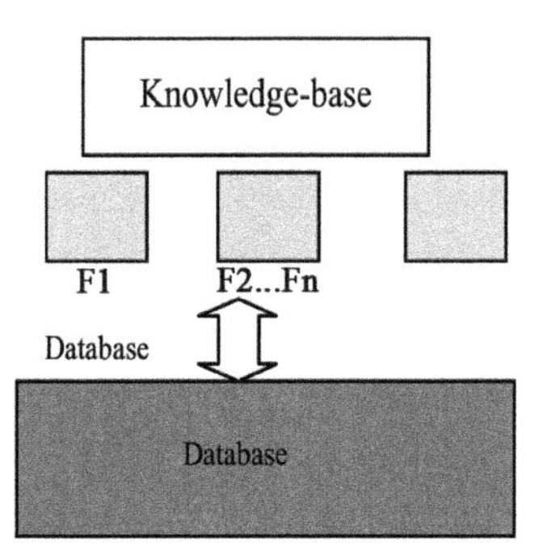

FIGURE 4.3 (See color insert.) Keyed data functions, inference, and model discovery.

Generalized predictive diagrams are defined, whereby specified diagram functions and search engine can select onto localized data fields. A Generalized Predictive Diagram is a predictive diagram where D(M) is defined from a minimal set of functions. The predictive diagram could be minimally represented by a set of functions {f1,…,fn} that inductively define the model. The functions are keyed onto the inference and knowledge base to select via the areas keyed to, designated as Si's in Figure 4.1 and data is retrieved Nourani (Nourani & Schulte, 2014). Visual object views to active databases might be designed with the above.

The trees defined by the notion of provability implied by the definition might consist of some extra Skolem functions {g1,…,gn}, that appear at free trees. The *f* terms and *g* terms, tree congruences, and predictive diagrams then characterize deduction with virtual trees Nourani (Nourani and Lauth, 2015) as intelligent predictive interfaces. Data discovery from KR on diagrams might be viewed as satisfying a goal by getting at relevant data which instantiates a goal. The goal formula states what relevant data is sought. We have presented planning techniques, which can be applied to implement discovery planning. In planning with G-diagrams that part of the plan that involves free Skolemized trees is carried along with the proof tree for a plan goal. The idea is that if the free proof tree is constructed then the plan has a model in which the goals are satisfied. The model is the initial model of the AI world for which the free Skolemized trees were constructed.

Partial deductions in this approach correspond to proof trees that have free Skolemized trees in their representation.

While doing proofs with free Skolemized trees we are facing proofs of the form p(g(.)) proves p(f(g(.)) and generalizations to p(f(x)) proves for all x, p(f(x)). Thus, the free proofs are in some sense an abstract counterpart of the SLD. Practical AI Goal Satisfaction. The predictive diagram could be minimally represented by a set of functions {f1,...,fn} that inductively define the model. The free trees we had defined by the notion of provability implied by the definition, could consist of some extra Skolem functions {g1...,gl} that appear at free trees. The *f* terms and *g* terms, tree congruences, and predictive diagrams then characterize partial deduction with free trees.

4.3.2 COMPETITIVE MODELS AND GOAL SATISFIABILITY

Consider an example ERP system with the goal to optimize a business plan with task assignments based on team-play compatibility (Figure 4.4).

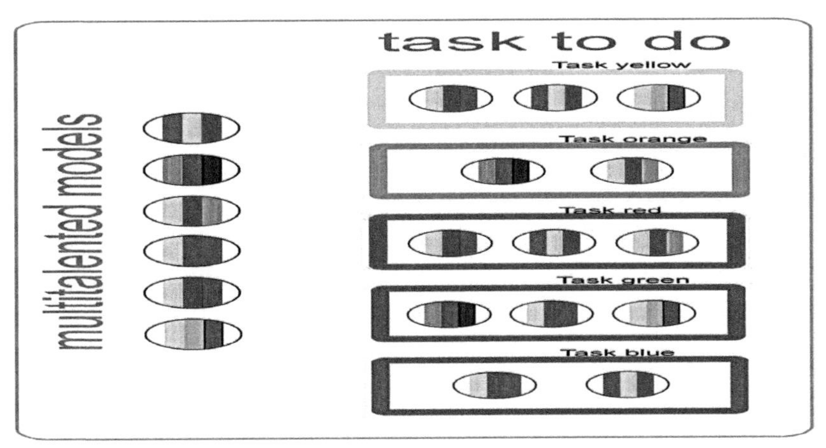

FIGURE 4.4 (See color insert.) Competitive models.

Generic model diagrams are basic function-based data modeling techniques the first author put forth over a decade ago to characterize a business domain, with business objects on a minimal function base [for example, Nourani, Grace & Loo, 2000].

Remark: The functions above are those by which a business model could be characterized with some schemes, e.g., above stock forecasting scheme example.

The computing specifics are based on creating models from generic model diagram functions where basic models can be piece-meal designed and diagrams completed starting from incomplete descriptions at times. Models uphold to a deductive closure of the axioms modeled and some rules of inference, depending on the theory. By the definition of a diagram, they are a set of atomic and negated atomic sentences. Thus, the diagram might be considered as a basis for a model, provided we can by algebraic extension, define the truth value of arbitrary formulas instantiated with arbitrary terms. Thus, all compound sentences build out of atomic sentences then could be assigned a truth-value, handing over a model. This will be made clearer in the following subsections.

4.4 DECISIONS, SPLITTING TREES, AND VECTOR SPANNING MODELS

Game theory is the study of rational behavior in situations in which choices have a mutual effect on one's business and the competitors. The best decision depends on what others do, and what others do may depend on what they think you do. Hence, games and decisions are intertwined. A second stage business plan needs to specify how to assign resources with respect to the decisions, ERP plans, and apply that to elect supply chain policies, which can in part specify how the business is to operate. The splitting agent decision trees have been developed independently by Nourani (1994). For example, when arranging team playing, there are many permutations on where the players are positioned. Every specific player arrangement is a competitive model. There is a specific arrangement that does best in a specific game. What model is best can be determined with agent player competitive model learning.

Cognitive agents are software agents included in the higher-level performance of autonomous intelligent systems. They belong to the class of autonomous agents which are complex computing entities active in some kind of environment without the direct intervention of humans or others virtual systems. In order to design intelligent agents systems, flexible problem-solving behavior and adequate knowledge about the beliefs regarding the environment and its changing conditions is required. We refer here to human, as well as to virtual systems and we try to look at some general characteristics of new cyber-physical systems. Multi-agent planning is modeled as a competitive learning problem where the agents compete on game trees as candidates to satisfy goals hence realizing specific models where the plan goals are satisfied (Figure 4.5). For example, agents can distinguish and priorities. The commitments are committed plans, committed goals, intended goals, and intended plans.

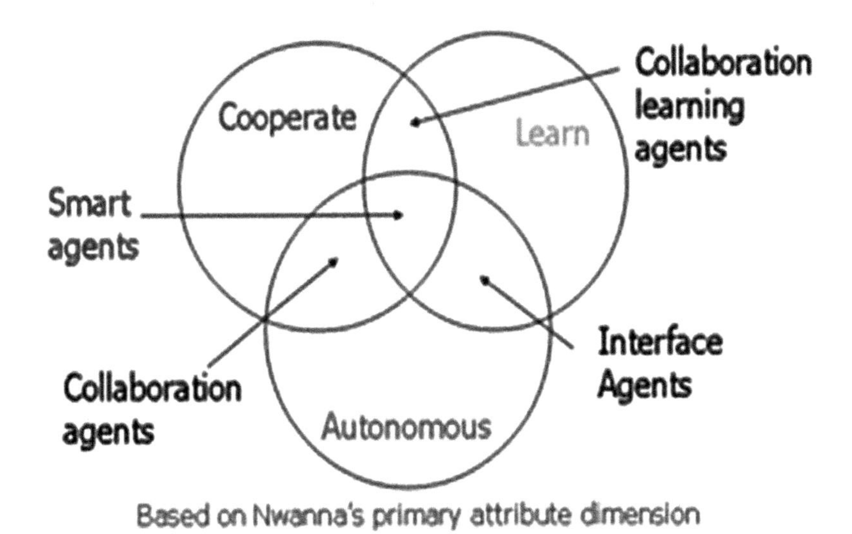

FIGURE 4.5 **(See color insert.)** Cognitive cooperative learning.

4.4.1 APPROACHES TO COGNITIVE SYSTEMS

Three broad approaches have been adapted so far for implementing cognitive concepts into autonomously acting systems: Data modeling approach that infers the cognitive concepts from the modular data structures to means-end reasoning system through a theorem prover. An example is a cognitive agent system called Artimis (Nourani et al., 2000), an intentional system designed for human interaction and applied in a spoken-dialog interface for human information access.

Procedural approaches that use explicit representations of cognitive contents approach based on the BID: belief, intention, desire agent models and can be instantiated in a procedural reasoning system (PRS), like for example, MARS (Nourani et al., 2000). Most cognitive systems fall into this second category. The newest approach is the “situated automata” approach that has no explicit representations of the cognitive concepts and therefore seems to perform better than other approaches in settings where higher performance is expected (Lauth, 2013).

Research into agent theories, architectures, and languages has a strong tradition in the agent systems field (see Nourani, 2013; Nourani and Lauth, 2013; Nourani, 1999b; Nourani, 1999; Nourani et al., 2000; Huhns & Singh, 1998). For an overview please look into the dissertation by Riemsdijk (2013). Cognitive agents are present in complex applications trying to solve efficiently: context-sensitive behavior, adaptive reasoning, ability to monitor and respond to the situation in real time (immersive agents), and modeling capabilities based on an understanding of human cognitive behavior, like innovation management, generation of new insights, e.g., new ways of thinking.

4.4.2 SPLITTING TREES AND THE CART MODEL

The following examples from Nourani (2010) can be motivating for business applications. Example: A business manager has 6 multitalented players, designed with personality codes indicated with codes on the following balls. The plan is to accomplish five tasks with persons with matching personality codes to the task, constituting a team. Team competitive models can be generated by comparing teams on specific assignments based on the task area strength. The optimality principles outlined in the first author's publications might be to accomplish the goal with as few a team grouping as possible, thereby minimizing costs.

Huhns and Singh (1998) had put forth the following section presents new agent game trees. Best decision depends on what others do, and what others do may depend on what they think you do. Hence, games and decisions are intertwined. A second stage business plan needs to specify how to assign resources with respect to the decisions, ERP plans, and apply that to elect supply chain policies, which can in part specify how the business is to operate. A tactical planning model that plans critical resources up to sales and delivery is a business planner's dream. Planning and tasking require the definition of their respective policies and processes; and the analyses of supply chain parameters. The above are the key elements of a game, anticipating behavior and acquiring an advantage. The players on the business planned must know their options, the incentives, and how do the competitors think.

Example premises: Strategic interactions.

Strategies: {Advertise, Do Not Advertise}

Payoffs: Companies' Profits Advertising costs $ million.

The AND vs. OR principle is carried out on the above trees with the System to design ERP systems and manage as Cause principle decisions. The agent business modeling techniques the author had introduced (Brazier et al., 1997; Lauth et al., 2012) apply the exact 'system as cause' and 'system as symptom' based on models (assumptions, values, etc.) and the 'system vs. symptom' principle via tracking systems behavior with cooperating computational agent trees. The design might apply agents splitting trees, where splitting trees is a well-known decision tree technique. Surrogate agents are applied to splitting trees. The technique is based on the first author's intelligent tree project (ECAI 1994 and European AI Communication journal) are based on agent splitting tree decisions like what is designed later on the CART system: The ordinary splitting tree decisions are regression-based, developed at Berkeley and Stanford (Brieman, Friedman et al., 1984; Breiman, 1996).

CART system deploys a binary recursive partitioning that for our system is applications for the agent AND/OR trees presented in Section 4.2. The term "binary" implies that each group is represented by a "node" in a decision tree, can only be split into two groups. Thus, each node can be split into two child nodes, in which case the original node is called a parent node. The term "recursive" refers to the fact that the binary partitioning process can be applied over and over again. Thus, each parent node can give rise to two child nodes and, in turn, each of these child nodes may themselves be split, forming additional children. The term "partitioning" refers to the fact that the dataset is split into sections or partitioned. CART trees are much simpler to interpret than the multivariate logistic regression model, making it more likely to be practical in a clinical setting. Secondly, the inherent "logic" in the tree is easily apparent.

The agent splitting decision trees have been developed independently by the first author since 1994. For new directions in forecasting and business planning (Nourani, 2002). Team coding example diagram from reach plan optimal games where a project is managed with a competitive optimality principle is to achieve the goals minimizing costs with the specific player code rule (Nourani et al., 2005).

4.4.3 DESIGNING A SPLITTING TREE DECISION SYSTEM

The BID-architectures upon which specifications for compositional multi-agent systems are based are the result of analysis of the tasks performed by individual agents and groups of agents. Task (de)compositions include specifications of interaction between subtasks at each level within a task (de) composition, making it possible to explicitly model tasks which entail interaction between agents. The formal compositional framework for modeling multi-agent tasks DESIRE is introduced here. The following aspects are modeled and specified:

(1) a task (de)composition;
(2) information exchange;
(3) sequencing of (sub)tasks;
(4) subtask delegation; and
(5) knowledge structures.

Information required/produced by a (sub)task is defined by input and output signatures of a component. The signatures used to name the information are defined in a predicate logic with a hierarchically ordered sort structure (order-sorted predicate logic). Units of information are represented by the ground atoms defined in the signature. The role information plays within reasoning is indicated by the level of an atom within a signature: different (meta) levels may be distinguished.

In a two-level situation, the lowest level is termed object-level information, and the second level meta-level information. Some specifics and a mathematical basis to such models with agent signatures might be obtained from Nourani (1998). The meta-level information contains information about object-level information and reasoning processes; for example, for which atoms the values are still unknown (epistemic information). Similarly, tasks that include reasoning about other tasks are modeled as meta-level tasks with respect to object-level tasks. Often more than two levels of information and reasoning occur, resulting in meta-information and reasoning. Information exchange between tasks is specified as information links between components. Each information link relates to output of one component to the input of another, by specifying which truth-value of a specific output atom is linked with which truth value of a specific input atom.

4.5 SPANNING AND THE MULTITIER MODEL

4.5.1 SPANNING ATTENTION ON DECISION TREES

When a specific agent group "wins" to satisfy a goal, the agent group is presenting a model consistent with an intended world model for that goal. For example, if there is a goal to put a spacecraft at a specific planet's orbit, there might be competing agents with alternate micro-plans to accomplish the goal (Rao & Georgeff, 1996). While the galaxy model is the same, the specific virtual worlds where a plan is carried out to accomplish a real goal at the galaxy via agents are not. Therefore, plan goal selections and objectives are based on the attention spans with competitive agent learning. This technique can be also used to solve highly interacting communication problems in a complex web application, web intelligence settings. The intelligent languages (Nourani, 1996; Pedersen et al., 2013) are ways to encode plans with agents and compare models on goal satisfaction to examine and predict via model diagrams why one plan is better than another, or how it could fail.

The state space agent modeling present techniques to span the attention state space, where specific agents with internal state set I can distinguish their membership. The agent can transit from each internal state to another in a single step. With our multi-board model (Nourani, 1998) agent actions are based on me and board observations. There is an external state set S, modulated to a set T of distinguishable subsets from the observation viewpoint. A sensory function s: $S \rightarrow T$ maps each state to the partition it belongs. Let A be a set of actions which can be performed by agents. A function action can be defined to characterize an agent activity action: $T \rightarrow A$. There is also a memory update function, mem: $I \times T \rightarrow I$.

State vector machines (SVM), agent vectors, and data mining referring to the sections above are applied to design cognitive spanning. SVM creates a set of hyperplanes in N-dimensional space that is used for classification, regression, and spanning in our approach. The SVM algorithms create the largest minimum distance to have competitive goals satisfied on models.

The BID-architectures upon which specifications for compositional multi-agent systems are based are the result of analysis of the tasks performed by individual agents and groups of agents. Task (de)compositions include specifications of interaction between subtasks at each level within a task (de) composition, making it possible to explicitly model tasks which entail interaction between agents. The formal compositional framework for modeling multi-agent tasks DESIRE is introduced here. The following aspects are modeled and specified: (1) a task (de)composition; (2) information exchange; (3) sequencing of (sub) tasks; (4) subtask delegation; and (5) knowledge structures.

Information required/produced by a (sub) task is defined by input and output signatures of a component. The signatures used to name the information are defined in a predicate logic with a hierarchically ordered sort structure (order-sorted predicate logic). Units of information are represented by the ground atoms defined in the signature. The role information plays within reasoning is indicated by the level of an atom within a signature: different (meta) levels may be distinguished. In a two-level situation, the lowest level is termed object-level information, and the second level meta-level information. Some specifics and a mathematical basis to such models with agent signatures might be obtained from Nourani (1998).

The meta-level information contains information about object-level information and reasoning processes; for example, for which atoms the values are still unknown (epistemic information). Similarly, tasks that include reasoning about other tasks are modeled as meta-level tasks with respect to object-level tasks. Often more than two levels of information and reasoning occur, resulting in meta-information and reasoning. Information exchange between tasks is specified as information links between components. Each information link relates to output of one component to the input of another, by specifying which truth-value of a specific output atom is linked with which truth value of a specific input atom. For a multi-agent object information exchange model (see, for example, Nourani, 1995).

4.5.2 MULTITIER BUSINESS MODELS

From Nourani (2012), the three Tier Designs are given in the box, for example.

Presentation Layer: Runs with the address space if one or more web servers

Business Logic Layer: Runs with the address space of one or more application servers.

Data Layer: Backend and database

The presentation layer contains components dealing with user interfaces and user interaction. For example, a visual JAVA standalone application.

A business logic layer contains components that work together to solve business problems. The components can be high-performance engines, for example, Catalog engines, typically written in Java or C++.

The data layer is used by the business logic layer to persist state permanently. Control of the data layer is one or more databases that home the standalone.

From an example content processing prototype (Nourani, 2007, 2013), we can glimpse on the applications for the above section. A basic keyed databases view provides the presentation of the model. It is the look of the presentation of the model, and it is the look of the application. We apply predefined user know functions on the view to present the applications look. The view should be notified when changes to the model.

The business logic updates the state of the model and helps control the flow of the application. With Struts, this is done with an Action class as a thin wrapper to the actual business logic. The model represents the two actual business state of the application. The business objects update the application state. The business objects update the Ap. ActionForm bean represents the Model state at a session or request level, bean represents the Model state at a session or request level, and not at a persistent level. The JSP file reads information from the ActionForm been using JSP tags. Our design applies the same functions that are presented on the view for specific application to generate a content model for specific applications.

4.5.3 BUSINESS PROCESS MODELING LANGUAGES

Business Process Model and Notation (BPMN) provides a graphical notation for specifying business processes in a *Business Process Diagram* (BPD) applies flowcharting similar to UML. One visualization option for agent spanning process spaces is to apply business process mapping to visualize what a business does by taking into account roles, responsibilities, and standards that can characterize business process management. By providing a notation that is intuitive to business users, we can specify with a language what complex process semantics are being carried out. The BPMN specification also provides a mapping between the graphics of the notation and the underlying constructs of execution languages, particularly Business Process Execution Language (BPEL). The primary goal of BPMN is to provide a standard notation readily understandable by all business stakeholders. These include the business analysts who create and refine the processes, the technical developers responsible for implementing them, and the business managers who monitor and manage them. Consequently, BPMN serves as a common language, bridging the communication gap between business process design and implementation.

Widespread adoption for such standard might help unify the expression of basic business process concepts. Our spanning techniques can be deployed with the above to modularize the processes.

4.5.4 SOCIAL MEDIA INNOVATIVE SYSTEMS APPLICATIONS

The goal of the "SOCIAL Serendipity Game" is to choose the Number One most supportive "SOCAP Bridger" from all teams and number one most supportive "SOCAP Bonder," and we let the teams choose on the second day these people and argue why they think that they chose Hero's brought the most to increase the quality of Social and Open Innovation for the success of one groups or all groups. The arguing part is triggering them into a new dimension on seeing the Hackaton as a Social (supportive) Innovation process, that they are starting creatively to analyze and also optimize it (over communication and interaction) this is the angels- devils game, but it helps to even increase communication as well as bonding and bridging. The arguing will be based on how the bridgers and the bonders are able to communicate their knowledge to the groups. We will not let them become personal or mean. Example application areas are being treated on Gatesense—an open innovation platform for designing new business solutions that have to deal with massive amounts of open data.

On the technical part, Gatesense is primarily an open IOT service platform for offering open data access, open data brokerage and processing of multiple data resources available in large cities and rural regions or in larger industry settings. From Nourani (2003) on multimedia database TAIM, we note that many data sets contain more than just a single type of data. The existing algorithms can usually only cope with a single type of data. How can we design methods that can take on multimedia data from multiple modalities? Are we to apply separate learning algorithms to each data modality, and combine their results, or are there to be algorithms that can handle multimedia multiple modalities on a feature-level. Learning causal relationships amongst visual stored data is another important area, which can benefit from our project. Most existing learning algorithms detect only correlations, but are unable to model causality and hence fail to predict the effect of external controls.

4.5.5 ADMISSIBLE HEURISTICS FOR PREDICTIVE MODELING

Nourani (1986) had developed free proof tree techniques since projects at TU Berlin, 1994. Free proof trees allow us to carry on Skolemized tress on game tree computing models, for example, that can have unassigned variables. The techniques allow us to carry on predictive model diagrams. Reverse Skolemization (Nourani, 1986) that can be carried on with generic model diagrams corresponding to what since Genesereth (2011) is applying on the game tree "stratified" recursion to check game tree computations. The free trees defined by the notion of provability implied by the definition, could consist of some extra Skolem functions{g1,...,gm}, that appear at free trees. The *f* terms and *g* terms, tree congruences, and predictive diagrams then characterize partial deduction with free trees. To compare recursive stratification on game trees on what Geneserth calls recursive stratification we carry on models that are recursive on generic diagram functions where goal satisfaction is realized on plans with free proof trees (Nourani, 1994–2007).

Thus, essentially the basic heuristics here is satisfying nodes on agent AND/OR game trees. The general heuristics to accomplish is a game tree deductive technique based on computing game tree unfoldings projected onto predictive model diagrams. The soundness and completeness of these techniques, e.g., heuristics as a computing logic are published since Nourani (1994) at several events, for example, AISB, 1995, and Systems and Cybernetics, 2005 (Nourani, 2015).

In computer science, specifically in algorithms related to pathfinding, a heuristic function is said to be admissible if it never overestimates the cost of reaching the goal, i.e., the cost it estimates to reach the goal is not higher than the lowest possible cost from the current point in the path (Breiman et al., 1984). An admissible heuristic is also known as an optimistic heuristic.

An admissible heuristic is used to estimate the cost of reaching the goal state in an informed search algorithm. In order for a heuristic to be admissible to the search problem, the estimated cost must always be lower than or equal to the actual cost of reaching the goal state. The search algorithm uses the admissible heuristic to find an estimated optimal path to the goal state from the current node. For example, in A* search the evaluation function (where/is the current node) is:

$$f(n) = g(n) + h(n)$$

where,

$f(n)$ = the evaluation function;

$g(n)$ = the cost from the start node to the current node;

$h(n)$ = estimated cost from the current node to goal; and

$h(n)$ is calculated using the heuristic function.

With a non-admissible heuristic, the A* algorithm could overlook the optimal solution to a search problem due to an overestimation in $f(n)$.

An admissible heuristic can be derived from a relaxed version of the problem, or by information from pattern databases that store exact solutions to subproblems of the problem, or by using inductive learning methods. Here, we apply the techniques on goal satisfiability on completive models briefed on the preceding section. While all consistent heuristics are admissible, not all admissible heuristics are consistent. For tree search problems, if an admissible heuristic is used, the A* search algorithm will never return a suboptimal goal node.

The heuristic nomenclature indicates that a heuristic function is called an admissible- heuristic if it never overestimates the cost of reaching the goal, i.e., the cost it estimates to reach the goal is not higher than the lowest possible cost from the current point in the path. An admissible heuristic is also known as an optimistic heuristic (Russell and Norvig, 2002). We shall use the term optimistic heuristic from now on to save ambiguity with admissible sets from mathematics (Shoenfield, 1967, or, first author's publications on descriptive computing, for the time being.

What is the cost estimate on satisfying a goal on an unfolding projection to model diagrams, for example with SLNDF, to satisfy a goal? Our heuristics are based on satisfying nondeterministic Skolemized trees. The heuristics aims to decrease the unknown assignments on the trees. Since at least one path on the tree must have all assignments defined to T, or F, and at most one such assignment closes the search, the "cost estimate," is no more than the lowest. Let us call that nomenclature as one compound word: "admissible heuristics," not to confuse that with the notion of admissible sets at mathematical logic in the first author's publications. We do not know if we can have a relationship to that mathematical area yet. To become more specific how game tree node degrees can be ranked, we state one example linear measure proposition. The Nourani-Schulte (2013) and big data heuristics applications were newer developments that showed how admissible heuristics are developed on our game decision trees.

Novel predictive modeling analytics techniques with decision trees on competitive models with big data heuristics are presented in brief in Nourani-Fähndrich (2016). Spanning trees are applied to focus on plan goal models that can be processed by splitting agent trees and vector spanning models. Sparse matrices enable efficient computability on big date heuristics. A brief on predictive analytics on "big data" are presented with new admissibility criteria. Tree computing grammar algebras semantic graphs are a basis for realizing tree goal planning examples.

4.6 CONCLUSIONS

New bases for splitting decision tree techniques for enterprise business systems planning and design with predictive models are developed. Decision systems and game tree applications to analytics towards designing cognitive social media business interfaces are presented. Planning and goal setting bases for decision tree satisfiability with competitive business models are coupled with predictive analytics models accomplish goals on multi-tier business systems. Spanning trees focus on plan goal models that can be processed on a vector state machine coupled with database preprocessor interfaces. Heuristics on predictive analytics are developed based on ranked game trees from the first authors preceding publications on economic games towards newer applications to decision tree heuristics. New business modeling languages for characterizing and visualizing the spanning techniques are being considered.

KEYWORDS

- **cognitive social media interfaces**
- **competitive model planning**
- **decision trees**
- **enterprise modeling**
- **heuristic analytics**
- **predictive analytics**

REFERENCES

Brazier, F. M. T., Dunin-Keplicz, B., Jennings, N. R., & Treur, J., (1997). DESIRE: Modeling multi-agent systems in a compositional formal framework, In: Huhns, M., Singh, M., (eds.), *International Journal of Cooperative Information Systems—Special Issue on Formal Methods in Cooperative Information Systems, 6*(1), 67–94.

Brazier, F. M. T., Jonker, C. M., & Treur, J., (1997). Formalization of a cooperation model based on joint intentions. In: *Lecture Notes in Computer Science,* Vol. 1193, pp. 141–155.

Brazier, F. M. T., Treur, J., Wijngaards, N. J. E., & Willems, M., (1995). Temporal semantics of complex reasoning tasks. In: Gaines, B. R., Musen, M. A., (eds.), *Proceedings of the 10th Banff Knowledge Acquisition for Knowledge-Based Systems Workshop*, KAW'95.

Breiman, L., Friedman, J. H., Olshen, R. A., & Stone, C. J., (1984). *Classification and Regression Trees.* Chapman & Hall (Wadsworth, Inc.): New York.

Bullard, J., & Duffy, J., (1998). A model of learning and emulation with artificial adaptive agents, *Journal of Economic Dynamics and Control, 22*(2), pp. 2, 179–207, Elsevier.

Codrina, L., Lasse, V., & Rasmus, B. (2013). Sead Bajrovic–Magdalena, Wiktoria Marcynova, Rasmus Uslev Pedersen-Gatesense – *Making Sense of Massive Open Data Resources into Sustainable Business Solutions.*

Codrina, L., Nourani, C. F., & Rasmus, U. P., (2013). Anna Blume CBS Copenhagen: Cognitive agent systems and intelligent systems interfaces design using human cognition. *Coauthors with Special Issue on Multidisciplinary Perspectives of Agent-based Systems.*

Genesereth, M. R., & Nilsson, N. J., (1987). *Logical Foundations of Artificial Intelligence.* Morgan-Kaufmann.

*Grosskopf, D.,&*Mathias, W., *(2009).The Process: Business Process Modeling Using BPMN. Meghan Kiffer Press.* ISBN978-0-929652-26-9.

Huhns, M., & Singh, M. P., (1998a). Cognitive Agents. In: *IEEE Internet Computing,* Nov–Dec 1998.

Huhns, M., & Singh, M. P., (1998b). Readings in Agents. Morgan, Kaufmann: San Francisco.

Kinny, D., Georgeff, M. P., & Rao, A. S., (1996). A methodology and technique for systems of BID agents. In: Van der Velde, W., & Perram, J. W., (eds.), *Agents Breaking Away, Proceedings of 7th European Workshop on Modeling Autonomous Agents in a Multi-Agent World* (Vol. 1038). MAAMAW'96, Lecture Notes in AI, Springer.

Lauth, C., (Blog). *What is the Multinova Approach?,* http://www.multinnova.com/blog/multinnova, (retrieved on 01/09/2013).

Lauth, C., Berendt, B., Pfleging, B., & Schmidt, A., (2012). Ubiquitous computing. In: Al Mehler, & Romary, L., (eds.), *Handbook of Technical Communication*, Walter de Gruyter.

LOFT, (2008). Logic, and the foundations of game and decision theory – 8th International Conference, Amsterdam, The Netherlands, *Series: Lecture Notes in Computer Science, Vol. 6006 Subseries: Lecture Notes in Artificial Intelligence* In: Bonanno, G., Löwe, B., Van der Hoek, W., (eds.), 1st edn., 2010, XI, 207 p, ISBN: 978-3-642-15163-7.

Miller, J. H., Butts, C. T., & Rode, D., (2002). Communication and cooperation, *Journal of Economic Behavior and Organization, 47*(2), 179–195.

Moore, R. C., (1980). Reasoning about knowledge and action, In: *AI Center Technical Note 191.* SRI International Menlo Park, California.

Nourani, C. F., (1984). Equation intensity, initial models, and AI reasoning, Technical Report, 1983, A: conceptual overview. In: *Proceedings Sixth European Conference in Artificial Intelligence*, Pisa, Italy, North Holland.

Nourani, C. F., (1991). Planning and plausible reasoning in artificial intelligence, diagrams, planning, and reasoning. *Proceedings of Scandinavian Conference on Artificial Intelligence*. Denmark, IOS Press.

Nourani, C. F., (1996). *Slalom Tree Computing – A Computing Theory for Artificial Intelligence* (Vol. 9, No. 4). Revised in A.I. Communication, IOS Press.

Nourani, C. F., (1997). *Intelligent Tree Computing, Decision Trees and Soft OOP Tree Computing*. Frontiers in soft computing and decision systems papers from the 1997 technical report FS-97-04 AAAI ISBN: 1-57735-0790 www.aaai.org/Press/Reports/Symposia/Fall/fs-97-04.html programming.

Nourani, C. F., (1998a). *Agent Computing and Intelligent Multimedia – Challenges and Opportunities for Business Commerce and MSIS*. Preliminary, University of Auckland Management Science.

Nourani, C. F., (1998b). *Business Modeling and Forecasting*. AIECAAAI99, Orlando, AAAI Press.

Nourani, C. F., (1998c). *Intelligent Languages – A Preliminary Syntactic Theory*. Mathematical foundations of computer science, Springer.

Nourani, C. F., (1999a). *Agent Computing, KB for Intelligent Forecasting, and Model Discovery for Knowledge Management*. AAAI Workshop on Agent-Based Systems in the Business Context Orlando, Florida. AAAI Press.

Nourani, C. F., (1999b). *Competitive Models and Game Tree Planning*. Applications to economic games. A version published at SSGRR, L'Auquila, Rome, Italy, Invited paper.

Nourani, C. F., (1999c). *Model Discovery Knowledge Management*. AAAI workshop on agent-based systems in the business context Orlando, Florida, July. AAAI Press, 51–92.

Nourani, C. F., (1999d). Business modeling and forecasting: Part I and II, *AAAI-Workshops on Agent-based Systems in E-Commerce*. Orlando Florida.

Nourani, C. F., (2000a). *Cyberspace Interfaces, Cyber Signatures, and Business*. WWW computing expert update-knowledge-based systems and applied, *AI, 3*(3), SGES Publications, Edinburgh University.

Nourani, C. F., (2002b). *Game Trees, Competitive Models, and ERP New Business Models and Enabling Technologies*. Management School, St Petersburg, Russia Keynote Address, Fraunhofer Institute for Open Communication Systems, Germany 11: 00-12: 00 www.math.spbu.ru/user/krivulin/Work2002/Workshop.html.

Nourani, C. F., (2002c). *Multiagent Games, Competitive Models, and Game Tree Planning Fall Symposium*. Intent Inference, AAAI, Boston, November AAAI Press.

Nourani, C. F., (2004). *Model Discovery, Intelligent W-Interfaces, and Business Intelligence with Multitier Designs*, CollECTeR LatAm 13–15, Santiago, Chile http://ing.utalca.cl/collecter/techsession.php.

Nourani, C. F., (2005a). *Business Planning and Cross-Organizational Models*, Workshop on Collaborative Cross-Organizational Process Design. Linz, Austria, http://www.mensch-und- computer.de/mc2005.

Nourani, C. F., (2005b). *Business Planning and Open Loop Control*, International Conference on Autonomous Systems, Tahiti, October 2005, IEEE Publications.

Nourani, C. F., (2005c). *Intelligent Multimedia Computer Science- Business Interfaces, Wireless Computing, Databases, and Data Mining*. American Scientific and Publishers, http://www.aspbs.com/multimedia.htmlDecember2004.

Nourani, C. F., (2013). *W-Interfaces, Business Intelligence, and Content Processing, Invited Industry Track Keynote*. IAT-Intelligent Agent Technology, Atlanta.

Nourani, C. F., (2015). A predictive tableaux visual analytics, Slovakia. *ICTIC Proceedings in Conference of Informatics and Management Sciences* (Vol. 4, Issue: 1). Publisher: EDIS Publishing Institution of the University of Zilina Powered by: Thomson, Slovakia. ISSN: 13399144, CD ROM ISSN: 1339231X ISBN: 9788055410029.

Nourani, C. F., & Codrina, L., (2013). *Rasmus Uslev Pedersen, Cognitive Agent Planning With Attention Spanning–IOT Special Session-IAT*, Atlanta.

Nourani, C. F., & Codrina, L., (2015). *Open Loop Control, Competitive Model Business Planning and Innovation ERP Communication, Management, and Information Technology ICCMIT'15* www.iccmit.net, Prague, Czeck Republic, April 2015. Proceedings Elsevier Publishers.

Nourani, C. F., & Grace, S. L., & Loo, K. R., (2000a). *Model Discovery From Active DB with Predictive Logic,* revised in June 1999, Data mining 2000 applications to business and finance, Cambridge University, UK.

Nourani, C. F., & Grace, S. L., & Loo, K. R., (2000b). *Revised June 1999, KR, and Model Discovery From Active DB with Predictive Logic Data Mining.* Applications to Business and Finance Cambridge University, UK, August 2000.

Nourani, C. F., & Hoppe, T., (1994). Intentional models and free proof trees. *Proceedings of the Berlin Logic Colloquium*. Humboldt University Mathematics.

Nourani, C. F., & Johannes, F. (2015). A formal approach to agent planning on inference trees Akdmkrd.Tripod.com. DAI TU Berlin NAEC 2014. Treiste.

Nourani, C. F., & Johannes, F., (2016). Decision trees, competitive models, and big data heuristics: DAI TU Berlin. *Written for Virtual Business Proceedings*. Slovakia, Thompson Publishers. Heuristics, Slovakia.

Nourani, C. F., & Maani, K., (2003). *Business Modeling and Enterprise Planning: Two Synergistic Partners SSGRR.* L'Auquila, Rome, Italy, Invited paper. University of Auckland, MSIS http://www.ssgrr.it/en/ssgrr2003w/papers.htm.

Nourani, C. F., & Oliver, S., (2013). *Multiagent Decision Trees, Competitive Models, and Goal Satisfiability*. DICTAP, Ostrava, Czeck Republic.

Nourani, C. F., & Oliver, S., (2014a). *Multiagent Games, Competitive Models, Descriptive Computing SFU*, Vancouver, Canada Draft, February 2011, Revision 1: September 25, 2012 SFU, Vancouver, Canada Computation Logic Lab. Europment, April, Prague.

Nourani, C. F., & Oliver, S., (2014b). Multiplayer games, competitive models, descriptive computing submitted for the EUROPMENT conferences EUROPMENT conference (Prague, Czech Republic, Wednesday 2, Thursday 3, Friday 4, and April 2014). *Computers, Automatic Control, Signal Processing, and Systems Science*, ISBN: 978-1-61804-233-0.

Nourani, C. F., Codrina, L., & Rasmus, U. P., (2002). Merging process oriented ontologies with a cognitive agent, planning, and attention spanning. *Submission Open Journal of Business Model Innovations,52.*

Nourani, C. F., Codrina, L., & Rasmus, U. P., (2013). *Agent Planning Models and Attention Span – A Preliminary*, IAT, Atlanta.

Nourani, C. F., Codrina, L., & Rasmus, U. P., (2015). CBS-merging process-oriented ontologies with a cognitive agent, planning, and attention spanning. *Open Journal of Business Model Innovations.*

Nourani, C. F., Grace, S. L., & Loo, K. R., (1999). *Intelligent Business Objects and the Agent Computing University of Auckland.* Parallel and distributed computing, Las Vegas, Nevada, CSREA Press www.informatik.uni- trier.de/~ley/db/indices/a- tree/l/Loo: Grace_SauLan. html.

Pedersen, R. U., (2010). Micro information systems and ubiquitous knowledge discovery In: *Lecture Notes in Computer Science*, No. 6202, 2010, 216–234.

Pedersen, R. U., Simon, J., Furtak H. I., Lauth, C., & Rob Van, K., (2013). Mini smart grid at Copenhagen business school: Prototype demonstration, In: Jan Vom, B., Riitta, H., Sudha, R., & Matti, R., (eds.), *Design Science at the Intersection of Physical and Virtual Design: 8th International Conference*. DESRIST 2013, Helsinki, Finland. Proceedings, Heidelberg, Springer, 446–447.

Rao, A. S., & Georgeff, M. P. (1996). Modeling rational agents within a BID-architecture, In: Fikes, R., & Sandewall, E., (eds.), *Proceedings of the Second Conference on Knowledge.*

Riemsdijk (2013). *CBS Copenhagen: Cognitive Agent Systems and Intelligent Systems Interfaces Design using Human Cognition.* Cyrus F. Nourani, Codrina Lauth, Rasmus Uslev Pedersen, Anna Blume. Special Issue on Multidisciplinary Perspectives of Agent-Based Systems.

Sieg, G., (2001). A political business cycle with bounded rational agents, *European Journal of Political Economy, Elsevier*, *17*(1), 3, 39–52.

Unin-Kęplicz, B., Verbrugge, R., & Slizak, M., (2010). TeamLog in action: a case study in teamwork. In: *Computer Science and Information Systems* (Vol. 8).

Vanhoof, K., & Lauth, I., (2010). Application challenges for ubiquitous knowledge discovery, In: *Ubiquitous Knowledge Discovery – Challenges, Techniques, and Application.* LNAI6202, State-of-art Survey, Springer, 108–125.

Von Neumann, & Morgenstern, O., (1994). *Theory of Games and Economic Behavior,* Princeton University Press, Princeton, NJ, p. 54.

Wooldridge, M., & Jennings, N. R. (1995). Agent theories, architectures, and languages: a survey. *Intelligent Agents – Intelligent Multimedia New Computing Techniques*, Design Paradigms, and Applications.

CHAPTER 5

A FRAMEWORK FOR DEVELOPMENT OF ONTOLOGY-BASED COMPETENCY MANAGEMENT SYSTEM

ILONA PAWEŁOSZEK

Częstochowa University of Technology, Faculty of Management, E-mail: ilona.paweloszek@wz.pcz.pl

ABSTRACT

Modern organizations are faced with rapid changes in technology and customer requirements. As a consequence, the need for new competencies emerges. In this situation, agile competency management is one of the most important drivers helping to foster a firm's competitive advantage. The competencies are complex in nature as they are composed of knowledge, skills, and abilities, which in turn can be considered at multiple levels of generalization. Therefore, competency management can be supported by semantic solutions exploiting domain competence ontology.

The aim of this chapter is to propose a semantic framework based on Software Competence Ontology, which can be used to support process-oriented competency management in a software development company. With this aim in mind, a solution was described which integrates business process models with competency information. The proposed Software Competence Ontology was designed using Protégé platform.

5.1 INTRODUCTION

The modern labor market is constantly changing due to continuous advancements in technology and innovative products. It is particularly vivid in IT companies, their services need to be more adaptable, agile, and flexible than ever before. Therefore, the need for new competencies continuously shows up and competencies that are already in existence, change their definitions (Różewski et al., 2013). This situation is reflected in a steady growth of interest in competency management systems.

Competency management is not new and has been widely practiced by all sort of companies that developed formal approaches to ensure that they have qualified human resources to meet their business goals. The practice of competency management consists in defining the skills and knowledge needed to run the business processes of an organization. Once they are defined, each employee or potential contractor should be described according to the formal definitions. The definitions can be used in many ways, such as forecasting and scheduling workforce, determine training goals and measure achievement of these goals.

Traditionally, competencies are based on functional role analysis (Keller, 2016). However, the popularization of process approach as a standard necessitates changing the ways of defining and identifying requisite competencies, which should be considered in the framework of tasks implemented within the process rather than by position of the employee in the organization hierarchy.

The aim of this chapter is to show how the ontology describing competencies can be linked with business process models to support process-oriented, dynamic competency management in a company. With this aim in mind, a proposal for practical implementation of ontology-enhanced business process model was presented and illustrated by the example of software development.

The concept of competence and the role of competence management systems have been described in Section 5.2. There are many taxonomies describing competencies, which can be potentially useful in different domains, a brief review of those frameworks are presented in Section 5.3. The Software Competence Ontology (SCO) developed upon the base of European e-Competence Framework is described in Section 5.4.

In the business process management area, the Business Process Modeling Notation (BPMN) is the de-facto standard approved by ISO/OSI. Section 5.5 explains the possibility to extend BPMN with user-defined elements. In this case, the added elements create the abstract layer describing competencies needed to perform the tasks included in the business process.

Section 5.6 presents an example of a process model enhanced by annotations on competencies needed to do particular tasks or groups of tasks within the process. The example process is software configuration management. The annotations contain references to the SCO. Connecting the business process with competence ontology allows for querying the database of process models, to find the set of competencies needed to run the process. It also allows for searching through the profiles of employees and job candidates to find the most appropriate team for process execution.

The last section outlines conclusions and directions for future discussions and development of the proposed framework.

5.2 THE NEED FOR THE SUPPORT OF COMPETENCY MANAGEMENT

Competencies are the combination of knowledge, skills, and abilities which are relevant in a particular job position and which, when acquired, allow a person to perform a task or function at a high level of proficiency (Osa, 2003). At the level of the individual, competencies are important for defining job or work content.

At the level of the organization, its core competencies are of great importance. This term was first used by C. K. Prahalad and Gary Hamel in 1990. It is a broad term and since then many definitions have been introduced. Prahalad and Hamel (1990) defined core competencies "as the collective learning of the organization, especially how to coordinate diverse production skills and integrate multiple streams of technology." The core competency of an organization can be defined as: "a harmonized combination of multiple resources and skills that distinguish a firm in the marketplace" (Schilling, 2013). Core competence is "what an organization does particularly well, in its purest sense it is a firm-specific collection of skills, insights, and capabilities that represent the product of long-term accumulated knowledge, organizational learning, and focused investment" (Kessler, 2013).

It, therefore, can be concluded that core competencies are usually a combination of abilities embedded in people, assets, and relations that result in the possibility to execute critical business processes at the highest level of competitiveness.

Competency management is an old, widely used practice that consists of all of a company's formal, organized approaches to ensuring that it has the human talents needed to meet its business goals. It is usually one of the roles of Human Resources department. Software systems support companies in gathering, storing, searching, and analyzing competency-related data. Manager dashboards provide quick insight of information across an entire company, and search functions help HR managers find out who has certain skills. Competency management system (CMS) is usually a part of the HR module of an enterprise system installed on the corporate server or accessible through the cloud to facilitate information sharing and cooperation among different departments. Competency management systems are also used for evaluating human resources and assessing job candidates. By integrating CMS with other workforce planning tools, such as scheduling and business process management systems, it is possible to identify the current and future competency gaps and surpluses. In the recent years, increased attention is being paid to improve and expand these systems (Maniu & Maniu, 2009). Many of the earliest efforts were custom developed by organizations such as NASA and the U.S. Coast Guard (Keston, 2013).

The existing competence management systems usually use databases and repositories of knowledge and skills descriptions, which are used in employee profiles to express levels and areas of their expertise. With the emergence of Web 2.0 after the year, 2001, enterprises started to use social networking tools to build and publish employee profiles. The mentioned solutions are very popular however they have two major drawbacks – lack of standardized vocabulary and lack of reasoning mechanisms to determine the competencies of individuals' by inferring on different levels of abstraction and detail.

These limitations are addressed by semantic solutions based on ontologies. For example, ontology-based skill management was proposed in the KOWIEN project carried out by L. Dittman (2004). HR ontologies integrating some of the existing standards and classifications with the aim of supporting recruitment process has been developed by M. Mochol et al. (2007) and A. Gomez-Perez et al. (2007). E. Biesalski and A. Abecker discussed the integration of HR processes with ontologies in a project at DaimlerChrysler AG to support strategic training planning (Biesalski & Abecker, 2005).

As it can be seen in recent years semantic models in form of ontology are an increasingly popular way of formalizing, encoding, and integrating knowledge from various sources to support business processes. In the domain of competency management, semantic models can provide common definitions that can facilitate information exchange within an enterprise and throughout an industry. Another advantage of semantics is the possibility of automatic reasoning on the basis of the semantic model and available information. This feature can be of great value while collecting data on competencies from many different resources such as internet portals with job offers, CVs of job applicants and databases of other organizations. Ultimately semantic models can help HR managers to compare and evaluate employees' knowledge and skills, which are useful while developing project staffing plans or employees' professional training.

There are many frameworks describing competencies or useful to define them, which can be potentially useful in different domains. These frameworks are briefly described in the next section.

5.3 SEMANTIC MODELS FOR COMPETENCY MANAGEMENT

A semantic model is a formal representation of domain knowledge. The semantic model consists of a network of concepts and the relationships between those concepts (Halper, 2007). Concepts area particular ideas or topics with which the user is concerned As such, a semantic model can take many forms, each of which has its own uses, and each of which is understandable to its end user. The most common forms of semantic models are glossaries of fact types and definitions, business object models, ontologies, or conceptual data models (Von Halle & Goldberg, 2009).

Competence models are one of the abstract layers of the information system. They describe the features and behaviors of people in relation to performed professional activities. Semantic models allow for organizing the area of competency management by unifying concepts, measures, and information resources about competencies. Therefore, the semantic competence models can be useful in many areas such as knowledge management, workload planning, career paths planning, personalizing vocational training, project management, periodic evaluation, employees' development.

From the point of view of an executive, an adequate estimation of employees efficiency is necessary, which can be presented as comprehend view of the knowledge, skills, and personality features with regard to already performed or potential tasks. In the context of IT support, the effective competency management requires first of all well-defined, formalized meaning and unified representation of competencies in perspective or business processes, finding balance between a level of detail in competency definition and complexity of management processes as well as technical background consisting of different systems and services for supporting the areas of human resources management, employee training and knowledge management.

Current activities in the area of modeling and standardization of competency management include a number of initiatives oriented on different applications, some of them are:

- IMS-RDCEO – The Reusable Definition of Competency or Educational Objective (RDCEO) specification provides a means to create common understandings of competencies that appear as part of a learning or career plan, as learning pre-requisites, or as learning outcomes. The information model in this specification can be used to exchange these definitions between learning systems, human resource systems, learning content, competency or skills repositories, and other relevant systems. RDCEO provides unique references to descriptions of competencies or objectives for inclusion in other information models (IMS, 2007).

- HR-XML is a library of XML schemas developed by the HR-XML Consortium, Inc. to support a variety of business processes related to human resources management. The competencies schema which is a part of HR-XML allows for capturing of information about evidence used to substantiate a competency together with ratings and weights that can be used to rank, compare, and evaluate the sufficiency or desirability of a competency (HR-XML, 2004).
- InLOC (2017) provides ways of representing intended learning outcomes, including knowledge, skills, and competencies, so that the related information may be communicated between and used by ICT tools and services of all kinds, interoperable.
- O*NET (2017) – is a database of all occupations in the US economy. It provides a taxonomy of competencies and their elements and such as knowledge, skills, abilities, and many other. The data was collected from companies operating in the United States. The O*NET database can also serve as a statistical tool to examine labor market in the USA because it contains results of measurement of competency levels.
- E-CF – European e-Competence Framework – provides a reference of 40 competencies as applied at the Information and Communication Technology (ICT) workplace, using a common language for competencies, skills, knowledge, and proficiency levels that can be understood across Europe (European e-Competence Framework, 2016).

The next section presents the proposition of Software Competence Ontology that can be used in numerous contexts in HR management of the software development company.

5.4 SOFTWARE COMPETENCE ONTOLOGY

Ontologies are the most powerful form of knowledge representation. They describe a piece of reality: a collection of classes of entities and the relationships among them. Ontologies contain a controlled vocabulary or terminology, plus a semantic network that encodes the relationships between each term of the vocabulary (Masseroli & Tagliasacchi, 2009).

Semantic networks are graph structures composed of a set of edges and nodes used to represent knowledge, in a particular domain. The nodes of the graph represent the domain concepts, and edges represent relations among them.

One of the major potential benefits of constructing ontologies is that they can be implemented in reasoning software, since by the analysis of the graph structure relations can be found that are not directly specified anywhere.

Codification of competencies is the area well-suited to exploit ontologies because competencies can be organized in a form of hierarchy that represents different levels of abstraction – from very general areas to very detailed pieces of knowledge and skills.

The first step of the codification process is creating or finding a proper taxonomy which encompasses all necessary terms and definitions. The next step is the identification of relationships between the concepts of the taxonomy. Then, the semantic model can be formally defined and presented as an ontology.

The ontologies of competencies may have a wide range of applications in information systems supporting decision taking in the domains of human resources, manufacturing, and other. Ontologies enhance searching capabilities of software applications. By using the logical structure of concepts and their relations it is possible to search people with similar or complementary competencies regarding the domain and level of knowledge and skills.

As a reference for formalizing and codifying employees' competencies, the European e-Competence Framework (E e-CF) was used. The choice was governed by many features of the E e-CF, which make it suitable for applying it in software development domain. The E e-CF is not based on job profiles but on particular knowledge and skills that can be needed in many positions. This approach is more flexible and suitable for project-oriented companies (which is common in the software industry) characterized by dynamic nature of the work environment, where employees are often from different departments and have different job titles. E e-CF provides a general and comprehensive specification of e-Competencies described in a multidimensional structure which consists of:

- 5 competence areas – the E e-CF distinguishes between five competence areas derived from the general framework of ICT business process consisting in five phases: (A) plan, (B) build, (C) run, (D) enable, (E) manage;
- 40 competencies;

- 5 proficiency levels, where 1 denotes the weakest knowledge or skills; and
- knowledge and skills examples.

All the dimensions can be adapted and customized into different contexts from ICT business. One drawback of the E e-CF is the lack of level specification for detailed knowledge and skills elements. There is only desired level assigned to competencies.

For example, the competence “A.1. IS and Business Strategy Alignment” has desired proficiency level 4 or 5 (in the 1–5 scale). Therefore, to compare the evidenced skills and knowledge of a person with the desired level of competency specified in the ontology it is necessary to make assumptions about the desired levels of knowledge and skills elements. The proposed ontology is to overcome this gap.

In the prototype solution described here, the assumption was made that the desired levels of skills and knowledge are inherited from competency class. Therefore, if the A1 Competence has a desired level 4 or 5, we assume that all the skills and knowledge elements also have the same desired levels (equal to or higher than 4).

The classes and relations defined in the Software Competence Ontology are presented in Figure 5.1.

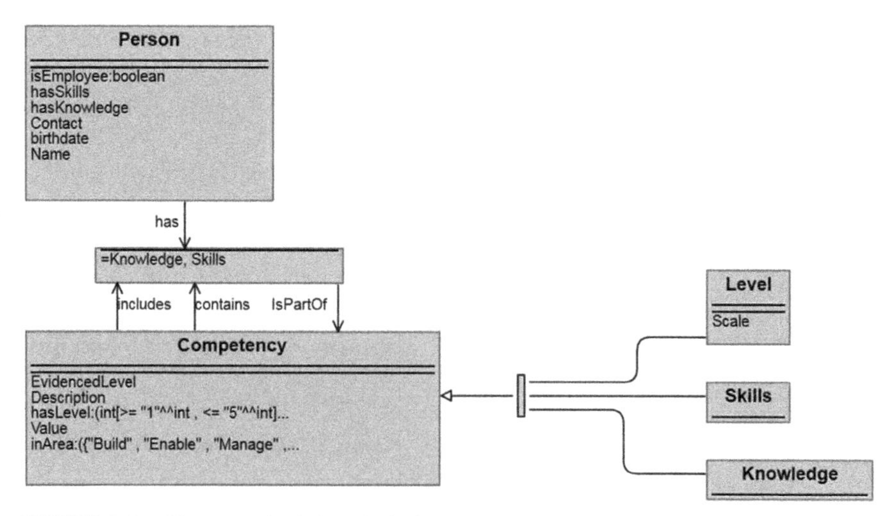

FIGURE 5.1 Classes and relations in Software Competence Ontology.

Source: Own elaboration, visualized in the OWLGrEd online tool: http://owlgred.lumii.lv

The proposed definition of classes, properties, and relations may facilitate many tasks from the area of human resources management. For example, by providing the possibility of semantic searching and reasoning the following queries can be posed:

- find persons who have evidenced level of competencies at least 3 in the area B (Build); and
- find a person who can substitute with manager X in the project Y.

The first example query is to analyze the profiles of employees and determine which of the evidenced competencies can belong to the area "Build" although the area may not be explicitly specified in their profiles.

In the second case, if the requirements for project Y are defined according to the areas and levels specified in the ontology, the aim is to find a person who at least fulfills these requirements (the levels of competencies of the person are equal or higher than requirements). Another approach is to find a person who is most similar to Employee X regarding the values of her competencies. Similarity can be calculated in many ways, e.g., applying selected distance measure and computing distance between levels of competence of each pair of the persons.

Because the competency values can be represented as a vector, cosine similarity measure can be used. The cosine similarity for two vectors A and B is calculated as follows:

$$Similarity = \cos(\theta) = \frac{A \cdot B}{\|A\| \cdot \|B\|} = \frac{\sum_{i=1}^{n} A_i B_i}{\sqrt{\sum_{i=1}^{n} A_i^2}\sqrt{\sum_{i=1}^{n} B_i^2}}$$

where:
A – the vector representing values of competencies of the employee X; and
B – the vector representing values of competencies of other employees.

The highest value of cosine similarity indicates the most appropriate (similar) candidate to substitute the Employee X.

The ontology of competencies also can be helpful if there is a need to analyze unstructured information such as CVs of candidates or new employees. In such a case semantic similarity measures can be used. The process of measuring semantic similarity is iterative and can be as follows:

1. First, the text of the CV is analyzed to find keywords that are present in the descriptions of ontology classes and properties.
2. The thesaurus is used to find similar words in CV to those that are present in the ontology.
3. Each time the keyword is found it is noted as one point for the given area and property of ontology.
4. When no more keywords are found the system displays suggestions of the areas and competencies identified for the given person.
5. The user engagement is needed to evaluate the competencies of the candidate in the scale 1–5 in the areas suggested by the system.

Semantic analysis of the terms used in descriptions of the employees' competencies is based on the use of a lexical database and semantic similarity algorithms. The lexical database WordNet (2010) can be used as it is particularly well suited for similarity measures, since it organizes nouns and verbs into hierarchies of is–a relation (Pedersen et al., 2004).

The Software Competence Ontology for the prototype of the system was coded using Protégé (2017) platform (Figure 5.2). The instances of competencies on the base of E e-CF were imported from MS Excel file using Cellfie plugin of the Protégé platform.

The ontology classes are displayed in a left-top window, on the left bottom of the figure there are instances of competencies. The "IS and Business Strategy Alignment" competence is selected. On the right top window knowledge and skills, elements for the selected competence are visible. Competencies are marked according to the 5 areas specified in the E e-CF framework and denoted with letters (A – Plan, B – Build, C – Run, D – Enable, E – Manage) and numbers.

The Software Competence Ontology can be used for many purposes which are summarized in Figure 5.3. First of all the ontology is to support annotation of documents evidencing competencies, such as descriptions in employee profiles and candidate CVs. The task of annotating is time-consuming if performed manually. Therefore, to facilitate the process of annotating the module of lexical analysis and annotation suggesting module can be used. Its role is to parse the documents and extract information on competencies, which are then compared with the SCO. Then, propositions of references to the ontology can be suggested. However, it should be noted

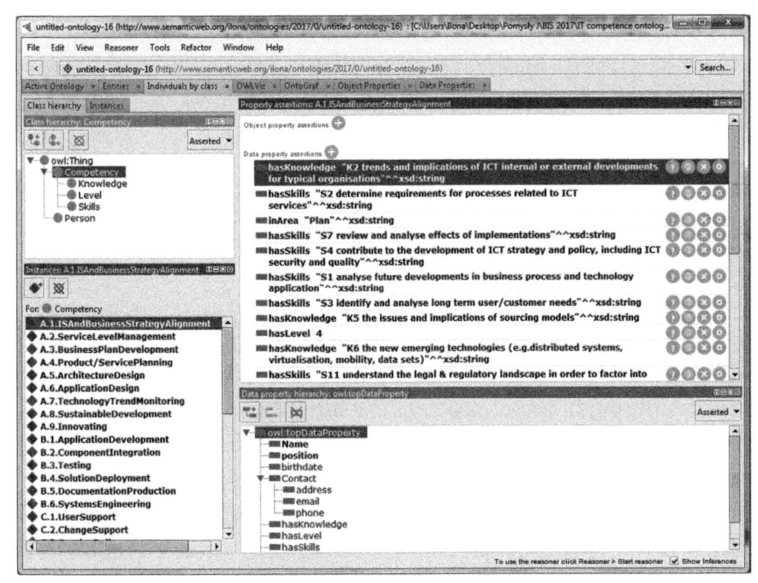

FIGURE 5.2 Software Competence Ontology displayed in protégé.

that the practical implementation of such a procedure would require additional dictionaries that contain synonymous words and phrases to those used in ontological descriptions of knowledge and skills.

Semantic searching can be performed on the base of annotated documents and business process models. Anytime the detailed knowledge and skills descriptions can be looked up in the SCO.

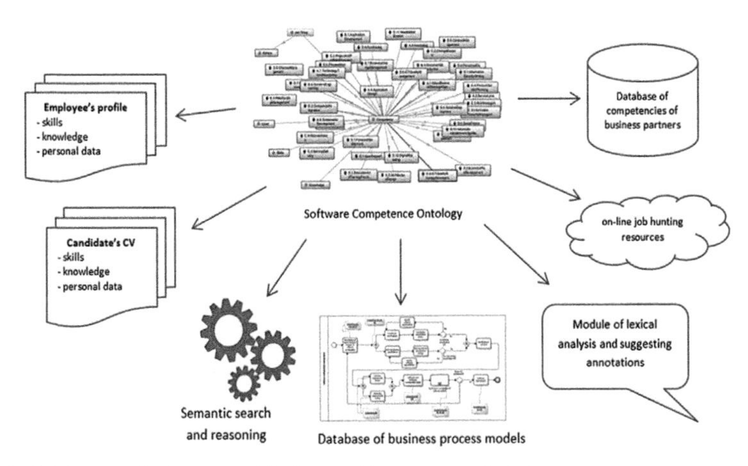

FIGURE 5.3 Possible usage of Software Competence Ontology.

The ontology can be also used by semantic search module that helps to find competency information in external online resources. However such resources cannot be annotated directly it is possible to analyze them on the fly and extract competency information.

5.5 COMPETENCY ANNOTATIONS FOR BUSINESS PROCESS MODELS

In the recent years, the concept of the process approach to management is gaining popularity among companies of all sizes. From the perspective of the process approach, the competencies should be seen in the framework of tasks implemented within the business process. Therefore, in the dynamic environment of contemporary organizations, a competency-oriented business process analysis (Leyking & Angeli, 2009) can be the right choice.

A Business Process Model is a step-by-step description of what one or more participants should do to accomplish a specific business goal. According to Gartner (2017) business process analysis tools are primarily intended for use by business end users looking to document, analyze, and streamline complex processes, thereby improving productivity, increasing quality, and becoming more agile and effective. Classic business process analysis is oriented toward analyzing and optimizing business processes for better productivity by saving time, costs or creating a more desirable product for customers.

Due to the increased need for agility competency-based management is the crucial activity of contemporary business organizations. In this situation, information technology support is the core element of the processes such as recruiting the most appropriate candidates, effective planning of employee development programs and project management. In many cases, the information about competencies is exchanged between collaborating organizations.

The IT tools for supporting process modeling should, therefore, provide the possibility to view the business process from the perspective of competencies required to perform particular tasks. Provision of information describing needed competencies of individuals involved in the process can help not only in workforce planning for the particular process but also in other management tasks such as expert finding, personalization of career paths and staff training.

The area of enhancing business process models by competency information requires resolving two basic issues. Unfortunately, none of the current business process modeling languages support the characterization of the business process in terms of competencies. Therefore, the first issue is to find appropriate notation to include competency data in process models. The second challenge is to design a formal representation that would be enough expressive to provide the detailed view of the process from the competency perspective.

In the business process management area, the Business Process Modeling Notation (BPMN) is the de-facto standard approved by ISO/OSI (1994) which allows for multi-view and a high-level description of business processes. BPMN provides a means to describe collaboration, choreography, and conversation aspects of business processes. However, it does not offer standard support for the characterization of the business process in terms of many other specific aspects. These aspects are often related to the area in which the process is executed, some formal regulations and standards that the process must comply with.

There are many attempts described in literature aiming at enhancing modeling notations by additional information, which would help to understand better the domain or offer the specific views of the process. For example, Rodríguez et al. (2012) propose an extension for including data quality requirements in process models. P. Bocciarelli and A. D'Ambrogio describe a BPMN extension for modeling nonfunctional properties of business processes (Bocciarelli & D'Ambrogio, 2011). The extension of BPMN facilitating security risk management was proposed by O. Altuhhova et al. (2013). Few works focus on the methodology of extending BPMN by user-defined elements (Stroppi et al., 2011), so it can be interpreted as a lack of maturity in this area.

BPMN2.0 offers an extensibility mechanism for enhancing standard BPMN notation with user-defined attributes and elements. This extensibility feature allows for the addition of new types of artifacts. Modeling tools may include features to hide or show these Artifacts. However the operations of adding the artifacts, hiding or showing them do not influence the sequence flow of the BPMN model. This is to ensure that BPMN diagrams always have a consistent structure and behavior (White & Miers, 2008).

The BPMN2.0 extension element consists essentially of four different classes which are (Lorenz et al., 2015):

- extension definition defines additional attributes;
- extension attribute definition presents the list of attributes that can be attached to any BPMN element; and
- extension Attribute Value contains attribute value.

The extension element of BPMN imports the definition and attributes with their values to the business process model.

Adding new concepts to the model provides the possibility to analyze it from different perspectives. From the point of view of competency management, BPMN models can be enhanced by artifacts representing the knowledge and skills necessary to run the process. Such an extension would allow for establishing and populating competence requirements across the organization, its business partners and job candidates. BPMN models with references to the descriptions of the required competencies create yet another perspective for analyzing the process performance regarding human factor. Moreover having a unified model for the description of competencies allows for addressing them on the stage of process design and further adjusting the process according to the current abilities of human resources.

In the next section, a proposal for practical implementation of ontology-enhanced business process model was presented and illustrated by the example of software configuration management.

5.6 GENERAL FRAMEWORK OF THE PROPOSED SOLUTION

This section describes a proposition of solution which integrates process models with competency ontology. The issue is presented on basis of software configuration management (SCM) process. SCM is one of the processes being an integral part of software engineering projects carried out by software companies. The process of software configuration management consists of identifying and defining the configuration items, controlling the release and change of these items throughout the system lifecycle, recording, and reporting the status of configuration items and change requests, and verifying the completeness and correctness of configuration items. It is a knowledge-intensive process that involves the cooperation of many participants such as, managers, analysts, developers, testers, and end-users.

The goal of the SCM process is to successfully deliver a software product to a customer or market in accordance with customer's requirements and software company's business plan (Farah, 2012). The decisions taken during this process by project managers are usually taken under the pressure of time and require skills from the areas such as software design, construction, testing, sustainment, quality, security, safety, measurement, and human-computer interaction.

The performance of the SCM process is essential for a software company because it directly impacts the customer satisfaction. Therefore, decisions taken during the SCM process should regard both the customer's needs and the business case perspective. Therefore, knowledge of related disciplines is very important as well as cognitive skills and behavioral attributes of the team members. The BPMN diagram of the software configuration management process enhanced by competence artifacts is illustrated in Figure 5.1.

The presented business process model is annotated by the information on needed competencies. The annotations are added as an additional artifact connected to the process tasks with a dotted line. To make the diagram more readable added elements contain symbols (for example, A1, B2, C2) which refer to the Software Competence Ontology – SCO (which is described later in this section). The symbols are displayed in form of hyperlinks so it is possible at any time to look up the detailed descriptions of knowledge and skills needed on each stage of the process.

In an example scenario, a project manager wants to find the right people for preparing software and hardware configuration report which is one of the tasks in software configuration management process (Figure 5.4).

The manager formulates a query to find persons who have competency denoted in the process model as: B3(S5), which is described in SCO as follows:

- B – area "Build" – consists of competencies needed for building software.
- B3 – Competence no.3: "Constructs and executes systematic test procedures for ICT systems or customer usability requirements to establish compliance with design specifications. Ensures that new or revised components or systems perform to expectation. Ensures meeting of internal, external, national, and international standards; including health and safety, usability, performance, reliability or compatibility. Produces documents and reports to evidence certification requirements."
- S5 – Skill no.5 of reporting and documenting tests and results.

The required level of competence is declared in the ontology using scale 1–5, where 5 means the most advanced knowledge and skills. Meanings of particular levels are also explained in the ontology. The classes and relations of the Software Competence Ontology are described in more detail in Section 4.3.

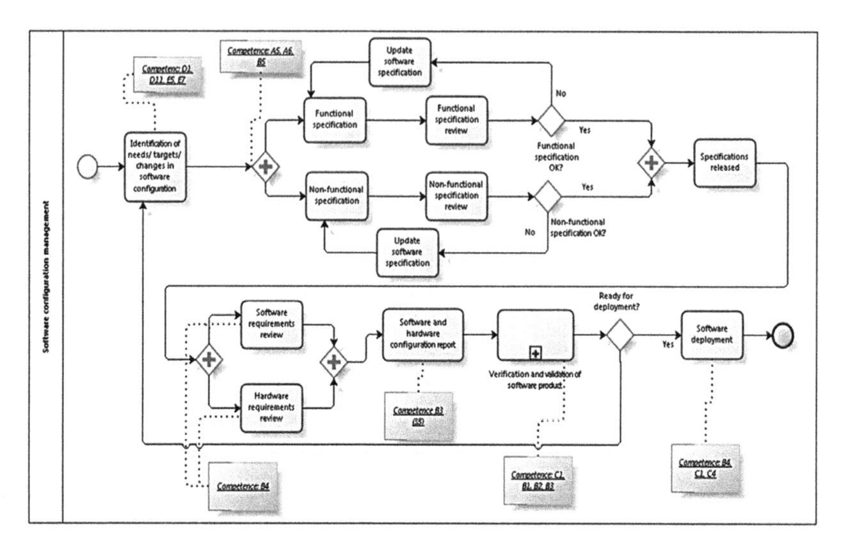

FIGURE 5.4 BPMN diagram of software configuration management process with competencies annotations. *Source*: (Pawełoszek, 2017).

The solution of integrating the previously presented process with Software Competence Ontology is aimed first of all to facilitate finding the right person to perform a task in the process. Moreover, the Software Competence Ontology can act as "common language" to describe the details of employees' profiles, annotating the CVs of job seekers, creating job postings and building a database of existing or potential business partners.

A general scheme of the platform supporting competency management is presented in Figure 5.5. The project manager or HR manager while analyzing the business process from the perspective of the competencies can formulate a query containing all the needed competencies and receive the list of potential contractors able to perform particular tasks.

The contractor can be an employee, a job seeker or a business partner who has knowledge and skills fully or partially consistent with the defined requirements. If there is no single person among the employees having all the required knowledge and skills for the given task, a team can be formulated consisting of the suggested individuals.

There are many possible data sources to use. The internal resources contain employees' profiles and CV of job candidates annotated with ontology concepts and references to knowledge and skills elements specified as the instances in the ontology. The external data sources may include information extracted from job-hunting websites and databases shared by

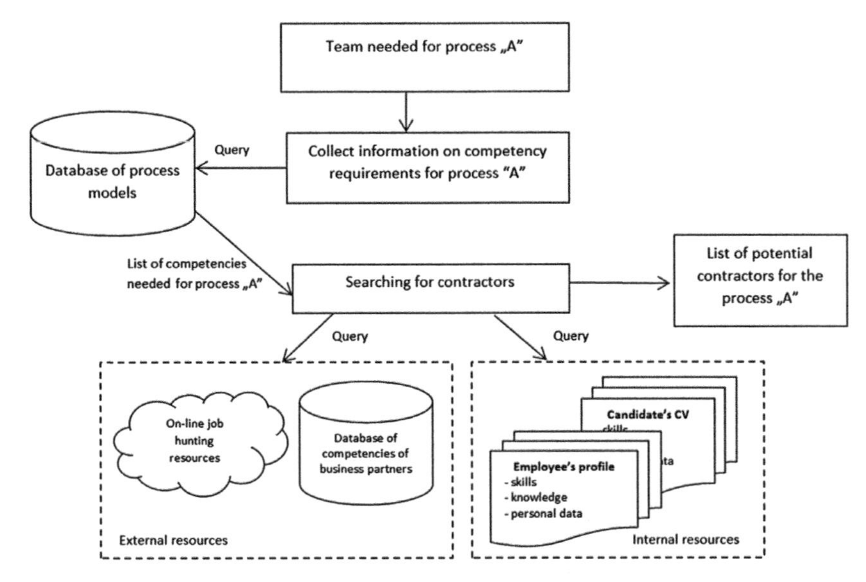

FIGURE 5.5 Usage scenario – workforce planning for a process.

other organizations. If there are no people with proper competencies among the employees the database of job candidates or external databases exposed by business partners or job-hunting portals can be searched through.

5.7 CONCLUSIONS AND FUTURE RESEARCH

The changes in technology and economic environment create the need for adjustment of business processes and searching or developing new competencies. Human capital is the carrier of the organization's knowledge and skills; therefore developing core competencies requires developing individual and team competencies.

The vision of the strategic core competencies is the driver for the development of collective or individual competencies. The requirements in this area should follow from the processes and tasks performed by the working group or the whole organization. Therefore, the main focus of this study is on individual and team competencies.

In this chapter, the concept of the platform for supporting process-oriented competency management has been proposed and illustrated by the example software configuration management process. The performance of the SCM process depends upon the proficiency of people involved in the process execution, therefore it is necessary to know the ideal bundle of competencies that the project team should evidence. Having a database of employees and job applicants profiles with specified skills, knowledge, and employment history, allows to match the competencies specified in the process model against evidenced levels of proficiency.

The dynamic approach to managing competencies on the base of business process flow can be valuable for process-based and virtual organizations where the environment is dynamic and frequent changes are needed to preserve the company's competitiveness and agility, this is often the case of software companies. IT labor market is, constantly changing due to continuous advancements in technology and innovative products. New competencies emerge and competencies that are already in existence, change their contents.

The Software Competence Ontology was developed in Protégé on the basis of European e-Competence Framework. The proposed Ontology will be further developed, especially by additional dictionaries and domain ontologies with the aim to improve semantic search algorithms. By the use of dictionaries of synonyms, it would be possible to enhance semantic search with the possibility of finding similar or the same competencies described with different terms. It is especially useful while dealing with external resources such as job hunters websites or databases of competencies that could be exposed by other business entities.

Competency management can be seen as one of the most important drivers of business processes performance improvement; therefore, there is a growing need for systematic approaches and IT support in this area.

ACKNOWLEDGMENT

This work was conducted using the Protégé resource, which is supported by grant GM10331601 from the National Institute of General Medical Sciences of the United States National Institutes of Health.

KEYWORDS

- **business process models**
- **semantic models for competency management**
- **Software Competence Ontology**

REFERENCES

Alluhhova, O., Matulevicius, R., & Ahmed, N., (2013). An extension of the business process model and notation for security risk management. *International Journal of Information System Modelling and Design, 4*(4), pp. 93–113.

Biesalski, H., & Abecker, A., (2005). *Human Resource Management with Ontologies.* Springer post-proceedings. Workshop on IT tools for knowledge management systems: Applicability, usability, and benefits (KMTOOLS), Springer, pp. 499–507.

Bocciarelli, P., & D'Ambrogio, A., (2011). A BPMN extension for modeling nonfunctional properties of business processes. In: *Proc. of the 2011 Symposium on Theory of Modeling & Simulation: DEVS Integrative M & S Symposium* (pp. 160—168). TMS-DEVS 2011, Society for Computer Simulation International, San Diego.

Dittmann, L., (2017). *Ontology-based Skills Management, Institute for Production and Industrial Information Management,* Essen 2004 https://www.pim.wiwi.uni-due.de/uploads/tx_itochairt3/publications/Arbeitsbericht_22.pdf (accessed September 23, 2017).

European e-Competence Framework: A common European framework for ICT Professionals in all industry sectors, (2016). http://www.ecompetences.eu/ (accessed September 23, 2017).

Farah, J., (2012). *Defining a Software Configuration Management Process to Improve Quality.* https://www.cmcrossroads.com/article/next-generation-process-and-quality–0 (accessed September 23, 2017).

Gomez-Perez, A., Ramirez, J., & Villazon-Terrazas, B., (2007). Reusing human resources management standards for employment services. In: *Proceedings of the Workshop on the First Industrial Results of Semantic Technologies (FIRST) at ISWC/ASWC* (pp. 28–41).

Halper, F., (2007). *What's a Semantic Data Model and Why Should We Care?* https://datamakesworld.com/2007/11/29/whats-a-semantic-model-and-why-should-we-care/ (accessed September 23, 2017).

HR-XML Consortium, Competencies (Measurable Characteristics) Recommendation, (2004). http://www.ec.tuwien.ac.at/~dorn/Courses/KM/Resources/hrxml/HR-XML–2_3/CPO/Competencies.html (accessed September 23, 2017).

http://protege.stanford.edu (accessed September 23, 2017).

http://www.gartner.com/it-glossary/bpa-business-process-analysis-tools/(accessedSeptember 23, 2017).

https://www.onetonline.org/ access: (accessed September 23, 2017).

IMS Global Learning Consortium: IMS Reusable Definition of Competency or Educational Objective Specification. https://www.imsglobal.org/competencies/ (accessed September 23, 2017).

InLOC, (2017). (Integrating Learning Outcomes and Competences) http://www.cetis.org.uk/inloc/Home (accessed September 23, 2017).

ISO 10303–203, (1994). Information technology – object management group business process model and notation.

Keller, S. J., (2016). *Competencies for the Community College Advisor: A Crucial Job in the Student Success Mission.* UNLV theses, dissertations, professional papers, and capstones. 2690. http://digitalscholarship.unlv.edu/thesesdissertations/2690.

Kessler E. H., (2013). *Encyclopedia of Management Theory.* Thousand Oaks, CA: Sage, p. 156.

Keston, G., (2013). *What is a Competency Management System?* http://www.kmworld.com/Articles/Editorial/What-Is-./What-is-a-Competency-Management-System–87130.aspx (accessed September 23, 2017).

Leyking, K., & Angeli, R., (2009). Model-based, competency-oriented business process analysis," *Enterprise Modeling and Information Systems Architecture Journal, 4*(1), pp. 14–25.

Lorenz, P., Cardoso, J., Maciaszek, L., & Van Sinderen, M., (2015). Software technologies – 10th International Joint Conference, ICSOFT 2015, Colmar, France. Revised selected papers," In: *Communications in Computer and Information Science* (Vol. 586, pp. 210–227). Springer.

Maniu, I., & Maniu. G., (2009). A human resource ontology for the recruitment process. *Review of General Management, 10*(2), 12–18.

Masseroli, M., & Tagliasacchi, M., (2009). Web resources for gene list analysis in biomedicine. In: Lazakidou, A., (ed.), *Web-Based Applications in Health Care and Biomedicine* (Vol. 7, pp. 117–141). Annals of Information Systems Series, Springer, Heidelberg.

Mochol, M., Wache, H., & Nixon, L., (2007). Improving the accuracy of job search with semantic techniques. In: *Proceedings of the 10th International Conference on Business Information Systems* (pp. 301–13). Poznan, Poland: Springer.

Osa, J. O., (2003). Managing the 21st-century reference department: Competencies. *The Reference Librarian, 39*, 35–50.

Pawełoszek, I., (2017). A process-oriented approach to competency management using ontologies. In: Ganzha, M., Maciaszek, L., & Paprzycki, M., (eds.), *Proceedings of the 2017 Federated Conference on Computer Science and Information Systems* (Vol. 11, pp. 1005–1013). ACSIS, http://dx.doi.org/10.15439/2017F441.

Pedersen, T., Patwardhan, S., & Michelizzi, J., (2004). WordNet: Similarity – measuring the relatedness of concepts. In: *AAAI 2004. San Jose*, CA (pp. 1024–1025).

Prahalad, C. K., & Hamel, G., (1990). The core competence of the corporation. *Harvard Business Review 68*(3), 79–91.

Princeton University "About WordNet." WordNet. Princeton University, (2010). http://wordnet.princeton.edu (accessed September 23, 2017).

Rodriguez, A., Caro, A., Cappiello, C., & Caballero, I., (2012). A BPMN extension for including data quality requirements in business process modeling. In: Mendling, J., & Weidlich, M., (eds.), *BPMN 2012. LNBIP* (Vol. 125, pp. 116–125). Springer, Heidelberg.

Różewski, P., Małachowski, B., & Danczura, P., (2013). The concept of competence management system for Polish National Qualification Framework in the Computer Science area. In: Ganzha, M., Maciaszek, L. A., & Paprzycki, M., (eds.), *FedCSIS* (pp. 759–765).

Schilling, M. A., (2013). *Strategic Management of Technological Innovation.* International Edition, McGraw-Hill Education, p. 117.

Stroppi, L. J. R., Chiotti, O., & Villarreal, P. D., (2011). Extending BPMN 2.0: Method and tool support. In: Dijkman, R., Hofstetter, J., & Koehler, J., (eds.), *BPMN 2011. LNBIP* (Vol. 95, pp. 59–73). Springer, Heidelberg.

Von Halle, B., & Goldberg, L., (2009). *The Decision Model: A Business Logic Framework Linking Business and Technology*. Taylor &: Francis, LLC, Auerbach. p. 336.

White, S., & Miers, D., (2008). *BPMN Modeling and Reference Guide: Understanding and Using BPMN.* Future Strategies Inc., Lighthouse Point, FL, USA.

CHAPTER 6

SUSTAINABLE MANUFACTURING AND SERVICES IN INDUSTRY 4.0: RESEARCH AGENDA AND DIRECTIONS

SUDHANSHU JOSHI[1], MANU SHARMA[1], RAJAT AGARWAL[2], and PANKAJ MADAN[3]

[1]*School of Management, Doon University, Uttarkhand, India*

[2]*Department of Management Studies, Indian Institute of Technology, Roorkee, India*

[3]*Faculty of Management Studies, Gurukula Kangri Vishvavidhaylaya, India*

ABSTRACT

The chapter aims to review literature in the area of Sustainable Manufacturing and Service Supply Chains in Industry 4.0 era. The survey aims to explore the literature growth on Industry 4.0, development of the theoretical framework and extending support to the practical implementation of Industry 4.0 in Service and Manufacturing setups. A noted list of 3,561 articles published during 1991–2018 are taken into consideration belong to various databases (viz. Web of Science and Scopus, that includes, Social Citation Index Expanded (SCI-Expanded), Social Science Citation Index (SSCI), Art and Humanities Citation Index (A&HCI), Conference Proceedings, etc.). The surveyed documents shows state of art research in the area of Industry 4.0 and industrial transformation across Service and Manufacturing setups.

We have varieties of Content Analysis Techniques, for present study Latent Semantic Analysis is used, due to its comprehensive nature and ability to a method to formulate latent factors (Joshi et al., 2017). LSA creates the scientific grounding to understand the trends. Using LSA, Understanding future research trends will assist researchers in the area of Manufacturing and Service Operations. Therefore, the study will be beneficial for practitioners of the strategic and operational aspects of technology adoption in Manufacturing and service supply chain decision making.

The current chapter uses the LSA approach (Joshi et al., 2017). The whole chapter is bifurcated into four sections viz. the first section describes the introduction to the evolvement of technology adoption across supply chains. The second section describes the usage of LSA for current study. The third section describes the finding and results. The fourth and final sections conclude the chapter with a brief discussion on research findings, its limitations, and the implications for future research. The outcomes of analysis presented in this chapter also provide opportunities for researchers/professionals to position their future technology driven supply chain research AND/OR implementation strategies.

6.1 INTRODUCTION

Last few decades witnesses transformation of industrialized economies into cyber-physical setups, aiming to meet customer needs (Oman et al., 2017; Santos et al., 2017; Sung, 2017; Zezulka et al., 2016; Zhong et al., 2017). The term Industry 4.0 is a comprehensive framework that creates 'punctuated equilibrium' and creates transformation in the traditional industries (Gould, 1989; Manhart, 2015; Siebel, 2017) and interconnect factors including, digitization, and integration of technical – economic matrix to complex technical – economical complex network; Digitization of Products and Services; Digitization of Business Models (Ahrens & Spöttl, 2015; Kagermann et al., 2015).

Virtually, a Virtuous Cycle has been established through Industry 4.0 and mass disruption and convergence between Science and Technology (S&T) and Social and Economic System leads to mass species extinction in both Manufacturing and Service Setups (Siebel, 2017). Also, the Triple Bottom Line (TBL) concept supports sustainable – Lean-innovative culture while creating a 'punctuated equilibrium' among People, Process, Policy, and Climate in both Manufacturing and Service Industry (Joshi and Sharma, 2014; Joshi and Sarkis, 2018). Functionally, both Service and Manufacturing operations aims to meet common aims of fulfilling customer demand through Information, Material, and Cash Flows (Chopra & Meindl, 2007). Industry 4.0, supports individualization of Products and Service at high flexible large-scale production, customization of services, sustainable partnerships and integrated value creation through high capacity of innovation, flexibility, and complexity (Rennung et al., 2016; Stock & Seliger, 2016; Joshi, 2018). In Industry 4.0 era, Supply Chain Sustainability becomes a focal area of discussion among practitioner and academicians.

The chapter incorporates extensive literature review in the area of Sustainable Supply Chains and Industry 4.0. The chapter aims to address two objectives. Firstly, extensive literature review, that aims to bridge gap between practice and theory. The secondly, the content analysis shall attempt to show multidisciplinary across area of research. The research approach has been adapted from Joshi et al. (2017).

To meet first objective, i.e., to review state-of-the-art cross-disciplinary relevance of the topic, 3561 research publications under various research disciplines (viz. Operation Research, Service Operations, Engineering, Business, and Economics, Big Data Analytics) are comprehensively examined from renowned resources including Web of Science (WoS), Scopus, IEEE Explore are evaluated.

Secondly, for content analysis, we used Latent Semantic Analysis (LSA). The rationale behind using LSA as analysis tool present research significantly contribute in three forms: (a) Conceptualization of research in the area of Sustainable Supply Chains and Industry 4.0, as the research roughly covers almost three decades of research (1991 till 2018); (b) Identification of emerging trends including new theories in Sustainable Supply Chains; (c) Assembling of multidisciplinary research under a common platform.

6.2 LATENT SEMANTIC ANALYSIS

LSA is an established and simplified bibliometric technique that explains the scope of research, its development in discipline of social science, computing, and life sciences (Deerwester et al., 1990; Landauer, 2007; Kulkarni, 2014; Joshi et al., 2017). Conceptually, LSA, blends varieties of theories from disciplines like linguistics, philosophy, computer science, blending of theories from philosophy, anthropology, linguistics, and psychology. LSA applications have witnessed recognition in variety of areas (Thorleuchter and Van den, 2014; Kwantes et al., 2016; Joshi et al., 2017). LSA is also used in Operation Management and Decision Sciences, based on semantic rules. The applications includes Business Intelligence, Strategic Decision making in Supply Chain (Hearst, 2000; Graesser et al., 1999; Wiemer-Hastings, Wiemer-Hastings, & Graesser, 2004).

Present study uses content analysis as quantitative assessment tool for textual analysis for the present research work on how different themes have emerged in the area of supply chain management in manufacturing and services, and Industry 4.0. The content analysis using LSA is based on grouping of keywords based on semantic searching from some database (viz. Scopus, Web of Science, IEEE Explore). LSA allows systematic analysis of corpus of literature related to some theme (Yalcinkaya and Singh, 2015). The present study focused on various aspects of Industry 4.0 applications including Smart Factory, Big Data Analytics, IoT, IoE, Smart Grid, Artificial Intelligence, and Digital-Twins, Elastic Cloud Computing in both service and Manufacturing Operations and Supply Chain Management. As mentioned in introduction, the present study aims fulfill the following objectives: (a) to address the theoretical development of subject i.e., Industry 4.0 and Sustainability in Manufacturing and Service Supply Chains; (b) to assess the emerging trends in the subject area; (c) to explore and illustrate scholarly development and future trends. During the primary pooling of Abstract we consider the fact that Industry 4.0" based knowledge is not only 'directly' accumulated from available data, rather also shows relation-based representation of the information content/blocks including (1) "Smart Factory" and "Sustainability" and "Supply Chain" and "Service Operations"; (2) "Service Operations" and "Sustainable development" and "Technology-Based Supply Chains"; (3) "Business" and "Supply Chain"; (4) "Economics" and "Supply Chain"; (5) "Industry 4.0" and "Cyber-Physical" and "Digital Twins"; (6) "Sustainability" and "Service Supply Chain" and "Digitization."

In order to do the comprehensive research, we have reviewed all cross-disciplinary areas which covered Operation Research, Service Operations, Engineering, Business, and Economics, Big Data Analytics, as mentioned in Joshi et al. (2017). Variety of publications was address for collection of articles including Scopus and Web of Science and also ACM, IEEE. We used LSA, based on its recent adoption in academic research (Landauer, 2007; Joshi et al., 2017). The scientific strength of LSA made it more authentic and acceptable as bibliometric analysis tool. The objective of using LSA is to process documents from the meta-dataset, using identified terms/keywords (Martin and Berry, 2007; Lisboa and Taktak, 2006; Evangelopoulos Zhang and Prybutok, 2012; Joshi et al., 2017). In the light of this situation, the book chapters deal in describing the semantically enhanced document retrieval system with an ontological multi-grained network of the extracted conceptualization.

6.2.1 VARIOUS STAGES OF LSA

LSA evolved since mid-1950s and later expanded into various domains (Lilley, 1954; Baker, 1962; Tarr and Borko, 1974, Van Rijsbergen, 1977;

Furnas, 1980; Furnas, 1985; Voorhees, 1985, Amsler, 1984; Bates, 1986; Deerwester et al., 1990). Research witnessed various applications of LSA from library indexing to business analytics (Kintsch, 2001; Jardin and Rijsbergen, 1971; Kundu et al., 2015; Landauer, 2007). The key objective behind using LSA over other content analysis techniques is to extract hidden knowledge from a set of text (Kundu et al., 2015) using extraction and querying large corpus of data (Deerwester et al., 1990; Han and Jiawei, 2007; Sidorova et al., 2008; Hossain, Prybutok, and Evangelopoulos, 2011). LSA process text[1] (a "document") from a set of files[2] (a corpus) and identified keywords (a "term"). Beside, LSA also help to identify latent factors[3] (a topic) from these extracted terms. Past research shows that the term Latent has got analogy from Human Mind (Valle-Lisboa and Mirzaji, 2007). Relevant challenges, issues, and applications of LSA (Evangelopoulos et al., 2012). Also, the mathematical modeling on LSA has been done (Martin and Berry, 2007; Valle-Lisboa and Mirzaji, 2007).

The standard process follows while conducting the Latent Semantic Analysis (LSA), as follows (Joshi et al., 2017):

Stage I: Vector Space Model (VSM): Corpus with d research documents is compiled (Salton, 1975; Porter, 1980). The result of this step will be generation of vocabulary of t number of terms. The researcher can eliminate common words from the list (viz. "the," "for," "of," "by," etc.).

Stage II: Matrix Generation: Term Frequency Matrix TFij is defined as the occurrence of the term (ith) in jth document. In order to normalize the frequency TF matrix shall be multiplied by inverse matrix (Kundu et al., 2015). The overall matrix is TF-IDF Matrix. Mathematically, TF-IDF Matrix can be expressed as:

$a_{ij} = f_{ij}.\log_2(N/n_i)$ (N = Total number of research documents, n_i = frequency of term i in the overall collection of research documents)

Stage III: Singular Value Decomposition (SVD): TF-IDF Matrix (Term Frequency-Inverse Matrix) has three matrices—U,S,V (U is the term eigenvectors; V is the document eigenvectors; S is the diagonal matrix of singular value). Over-fitting of factors needs to be avoided.

---

[1] Process Text is known as document in LSA Terminology.

[2] Set of Files is known as Corpus in LSA Terminology.

[3] Latent Topic is known as Topic in LSA Terminology.

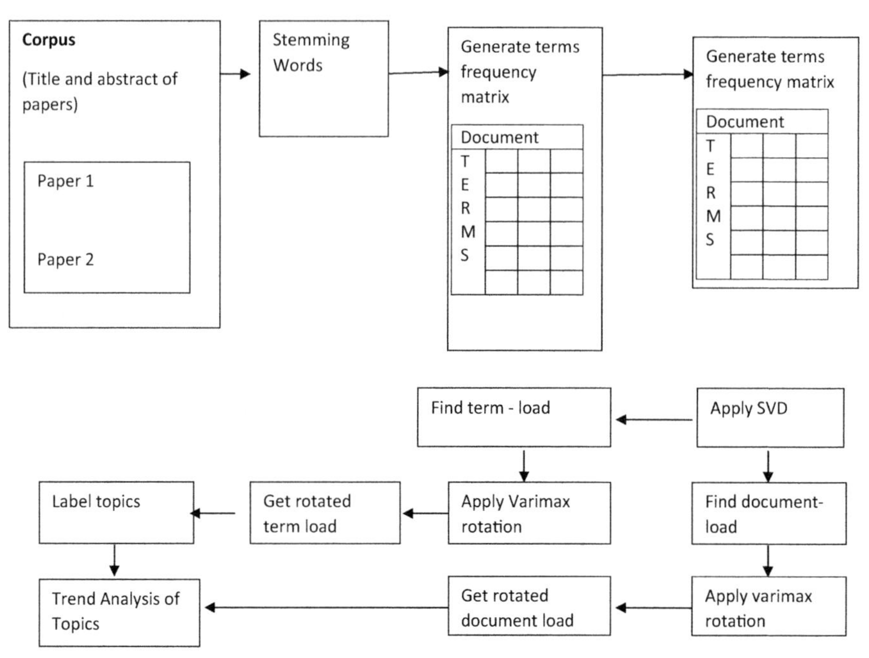

FIGURE 7.1 Flowchart of the data collection and analysis (adapted from Kundu et al., 2015)

(Reprinted with permission from Kundu, A., Jain, V., Kumar, S., & Chandra, C. (2015). A journey from normative to behavioral operations in supply chain management: A review using Latent Semantic Analysis. *Expert Systems with Applications*, 42(2), 796–809. © 2015 Elsevier.)

Kundu et al. (2015) has demonstrated the process of loading of term and document for each factor. This method is known as varimax. The aim of this method is to simplifying the term (i.e., High Loading terms using UkSk and high loading documents using VkSk matrics. These metrics correlate latent terms to particular factors.

After successful sample text mining, the final LSA has been performed on all abstract published with the keywords on "Sustainable Supply Chain Management" and "Industry 4.0" across noted list of articles. Exhibits below shows the documents trends in selected set of publications.

EXHIBIT A: Sample Research Documents for LSA Study

Document ID	Document Title	Procedia Engineering and Energy Procedia (Volume 141 and Volume 210)
RESDOC 1	• Problem identification and searches for optimized solution	43–52 and 32–37
RESDOC 2	• Models for problems related to Industry 4.0, Sustainability, and Supply Chain Management	25–29 and 114–132
RESDOC 3	• Clustering data that are graph connected	89–97 and 34–59
RESDOC 4	An algorithmic framework for the exact solution of three-star problems	54–66 and 72–87
RESDOC 5	• Proactive solution to the scheduling problems	67–74 and 87–95
RESDOC 6	• An iterated greedy heuristic for a market segmentation problem with multiple attributes	75–87 and 92–101
RESDOC 7	• Risk-cost optimization for overall planning in multi-tier supply chain by Pareto Local Search with relaxed acceptance criterion	88–96 and 32–49

TABLE B: Frequency Matrix1 (7X7)

Term	RESDOC 1	RESDOC 2	RESDOC 3	RESDOC 4	RESDOC 5	RESDOC 6	RESDOC 7
Research Approach	1	0	0	0	0	0	0
Sustainability Issues	1	1	0	1	1	1	0
Behavioral Models	0	1	0	0	0	0	0
Clustering	0	0	1	0	0	0	0
Sustainable Solutions	0	0	0	1	1	0	0
Triple Bottom Line (TBL) Optimization	0	0	0	0	0	0	1
Technology Adoption	0	0	0	0	0	0	1

EXHIBIT C: Frequency Matrix2 (7X7)

Term	RESDOC 1	RESDOC 2	RESDOC 3	RESDOC 4	RESDOC 5	RESDOC 6	RESDOC 7
Research Approach	1.658	0	0	0	0	0	0
Sustainability Issues	1.658	1	0	1	1	1	0
Behavioral Models	0	1.658	0	0	0	0	0
Clustering	0	0	1.658	0	0	0	0
Sustainable Solutions	0	0	0	1.658	1	0	0
Triple Bottom Line (TBL) Optimization	0	0	0	0	0	0	1
Technology Adoption	0	0	0	0	0	0	1.658

EXHIBIT D: Term Loading

	Before Varimax rotation		After Varimax rotation	
Term	**Factor A**	**Factor B**	**Factor A**	**Factor B**
Research Approach	0.322	–0.25	0.875	–1.32
Sustainability Issues	0.322	–0.25	0.875	–1.32
Behavioral Models	0.427	0.051	1.642	0.143
Clustering	0.299	0.353	1.149	0.984
Sustainable Solutions	0.729	0.330	2.801	0.918
Triple Bottom Line (TBL) Optimization	0.300	–0.387	1.154	–1.222
Technology Adoption	0.228	–0.553	0.875	–1.539

EXHIBIT E: Document Loading

	Before Varimax rotation		After Varimax rotation	
Term	**Factor A**	**Factor B**	**Factor A**	**Factor B**
Research Approach	0.1886	–0.6306	0.733	–1.539
Sustainability Issues	0.6013	0.4194	2.309	–1.539
Behavioral Models	0.1115	0.0185	0.428	0.143
Clustering	0.1236	0.2017	0.475	0.984
Sustainable Solutions	0.5772	–0.5395	2	0.918
Triple Bottom Line (TBL) Optimization	0.4915	0.3072	1.154	–1.079
Technology Adoption	0.1886	–0.6306	0.875	–1.539

6.3 DATA COLLECTION

Based on Journal Citation Report (JCR), top two journals on Operation Research and Operation Management (OR/OM) journals (1991–2018) with highest citation articles are selected as source of secondary data for the present study.

The rationale behind the selection of 'Energy Procedia' and 'Procedia Engineering' for LSA study, because of two major reasons: (a) The journals match the subject, objectives, and scope of the Study; (b) Collectively, journals generate a larger pool of cited research papers published and being one of the oldest Journal in the Operations Research and Management Sciences.

Both journals has same initial publication year, i.e., 2009 and has completed 8 years anniversary in 2017. Energy Procedia and Procedia Engineering carries cites core 1.16 and 0.74, respectively. The journals is also affiliated with Leading Institutions in the area of Operations and Service Supply Chain including INFORMS and other IFORS-affiliated or societies. The journal is indexed in Scopus, Science Citation Index Expanded, INSPEC, International Abstracts in Operations Research, etc. The study incorporate more than 15,000 articles published in both journals since 2009. The title and abstracts of research papers are taken from electronic databases including Web of Knowledge, Science Direct and Scopus. We have only considered journal articles based on selection criteria 'Industry 4.0' and 'Sustainable Supply Chain Management' as search keywords, broadly from 1991–2017. Both journals qualified the criteria and come under the broader framed initial timeline. In year 2017 alone, there exist 9,811 documents and their abstracts match the initial criteria out of total publications 14,429. Table 6.1 depicts citation Summary of both Journals

TABLE 6.1 Citation Summary

Results found	153,138
Sum of the Times Cited	18,776
Sum of Times Cited without self-citations	17,766
Citing Articles	76,329
Citing Articles without self-citations	91,207
Average Citations per Item	19.50
h-index	154

The overall sum of times cited of all research documents are 17,700. Sum of times cited without self-citations are 91,207. Average citation per research article was found 19.50 and the overall h-index was 154. Top 100 cited research articles with the components of "Industry 4.0" and "Supply Chain Management" are examined.

6.4 PREPARATION OF CORPUS

The number of articles published in Energy Procedia and Procedia Engineering from 2009 till 2017 which are sub-classified into Set 1/and Set 3 (2009–13), Set 2/and Set 4 (2014–17) for both Journals. We have classified the high volume set of documents including four sets/corpus viz. Set 1, Set 2, Set 3, and Set 4, respectively.

6.4.1 GENERATION OF THE TERM FREQUENCY MATRIX

The following steps have been done for each corpus – Set 1, Set 2, Set 3, and Set 4. SMART (System for the Mechanical Analysis and Retrieval of Text) is being used as the software to perform 'Text Stemming' (Deerwester et al., 1990) (adapted from earlier studies: Robertson, 2004; Salton, 1975, 1975a; Joshi et al., 2017). Term stemming is done and terms with a single occurrence in the corpus were excluded. This step produced a vocabulary of stemmed terms. Term frequency matrix is transformed to term frequency-inverse document frequency (TF–IDF) matrix to compress the frequencies. This is achieved by multiplying the Term Frequency (TF) by the Inverse Document Frequency (IDF). Generalized expression for each element of TF–IDF matrix is:

$$Wij = Tfij. IDFi = Tfij. \log 2 (N/ni)$$

where, N is the total number of documents in the collection and n_i is the frequency of term i in the entire collection of documents. A Singular Value Decomposition (SVD) operation is applied on TF–IDF matrix to get latent factors. List of Topics/Latent Factors are mentioned in Table 6.2. We generate term load and document load matrix (using varimax method) to find the best correlation of a term (or document in document load) with a latent factor. Results for four different corpus sets are presented in the next section. Factor labeling is achieved with the help of expert opinion (based on term load values). While preparing the corpus, we have used a 'year-documentID' (such as 2011-xxxxxx) format to create document file names (Kundu et al., 2015; Joshi et al., 2017). It is useful to utilize document-load to find a time trend of research direction.

6.4.2 TOPICS/LATENT FACTORS IDENTIFICATION

Based on the classified set of abstract several Latent Factors are being identified (Table 6.2).

6.5 RESULTS AND DISCUSSION

The results are illustrated in the form of graphs and tabulated charts to enhance the readability of the reports. First, we will present the outcome

TABLE 6.2 Latent Factors Identified Across Publications on 'Industry 4.0' and 'Sustainable Supply Chain Management'

Title	Topic/ Latent Factor	Theory/ Method Used
1.1	Factor 0: System View of Industry 4.0	Theoretical Framework, Absorption Theory, Theory of Constraints (ToC)
1.2	Factor 1: Supply Chain Competency	Content Analysis, Neo-Institutional Theory
1.3	Factor 2: Service Delivery Mechanism	Multi Criteria Decision Making (MCDM) Tools, Fuzzy AHP
1.4	Factor 3: Process Design	Fuzzy Sets, Conjoint Measurement
1.5	Factor 4: Strategic Design	Maintenance policies, Multi-Criteria Decision Making (MCDM)
1.6	Factor 5: Organization Culture, Human Resources and Decision Making	Data Envelopment Analysis (DEA), Russell Measure, Translation-Invariance
1.7	Factor 6: Service Distribution Planning	LPP, Network Analysis and Graph Theory
1.8	Factor 7: Information Management	Stochastic Model, Big Data Security, Data Mining
1.9	Factor 8: Technology Adoption	Technology Adoption Model (TAM), Technology Adoption Technique (TAT), UTAUT
1.10	Factor 9: Industrial Analysis	
2.1	Factor 10: Integration between Service Operations and Supply Chain Systems for Overall Performance improvement	Performance Evaluation Frameworks, System Theory
2.2	Factor 2: Operation Inventory Management	Linear- Programming Approach, Tabu Search Algorithm, Heuristic Algorithms, Genetic Algorithms
2.3	Factor 3: Competitive Positioning of the Service Offered	Game Theory, Theory of Constraints
2.4	Factor 4: Service- Product continuum	Weighted Completion-Time, Complexity Analysis
2.5	Factor 5: Supply Chain Decision Making	Game Theory
2.6	Factor 7: Logistics in Service Operations	Data Envelopment Analysis, Programming Approach, Tabu Search Algorithm, risk-based algorithm, branch, and bound, hill-climbing Algorithm
2.7	Factor 8: Supplier Relationship Management	Game Theory
2.8	Factor 9: Greening the Industry 4.0	

TABLE 6.2 *(Continued)*

Title	Topic/ Latent Factor	Theory/ Method Used
2.9	Factor 10: Information Sharing and Security among Supply Chain Partners	TAT, TAM, Game Theory
2.10	Factor 11: Service Delivery and Product/ Service Redesign	Siting Models, Heuristic Concentration, Algorithmic Approach, Weber Problem
3.1	Factor 0: Process Optimization for Service Supply Chains	Control Theory, Stochastic Modeling, Game Theory, Markov Process, Mixed Integer Programming, Simulations Techniques
3.2	Factor 2: Closed Loop Supply Chain in Service Industry	Storage- Retrieval System, Traveling Salesman Problem, Branch, and Bound, Neighborhood Search, Expectation Maximization, Problem Specific Solution
3.3	Factor 3: Inter-Firm Service Process Optimization	Hypothetical Testing, Survey
3.4	Factor 4: Pricing Decision	
3.5	Factor 5: Supply Chain Decision Making	Plant location Problem, Heuristic Concentration
3.6	Factor 6: Service Supply Chain Forecasting	Mathematical Programming, Algorithm, UML
3.7	Factor 7: Order Management	Generic Algorithm, Heuristics
3.8	Factor 8: Revenue Sharing Model	Bargaining Model, Risk –aversion, Cooperative game theory
3.9	Factor 9: Service Operations	Combinational Optimization
3.10	Factor 10: Supplier –Buyer Relation	Multi-criteria Analysis, AHP, Vector-Space Formulation

(Framework adapted from Joshi et al., 2017).

of LSA, identified as the top 10 topics (or latent factors) in an assembled format over approximately 9 years of both journals, 'Energy Procedia' and 'Procedia Engineering.' Next is the disassembled representation of the trends in topics of 'Industry 4.0' and 'Supply Chain Management and Operations Management.' We have grouped the Latent Factors into three groups and 10 prominent/latent factors each in group, i.e., overall 30 latent factors were analyzed (adapted from Joshi et al., 2017).

To research trends based on the latent factors can be acknowledge from various graphs in Graph 1 subcategory. Where distribution across normalized frequency of publication can be understand across last 9 years based on Core Topics/Latent Factors identified.

6.6 TREND ANALYSIS

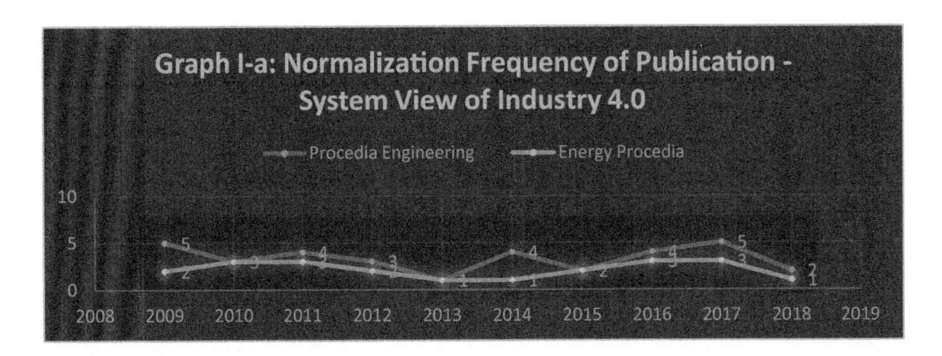

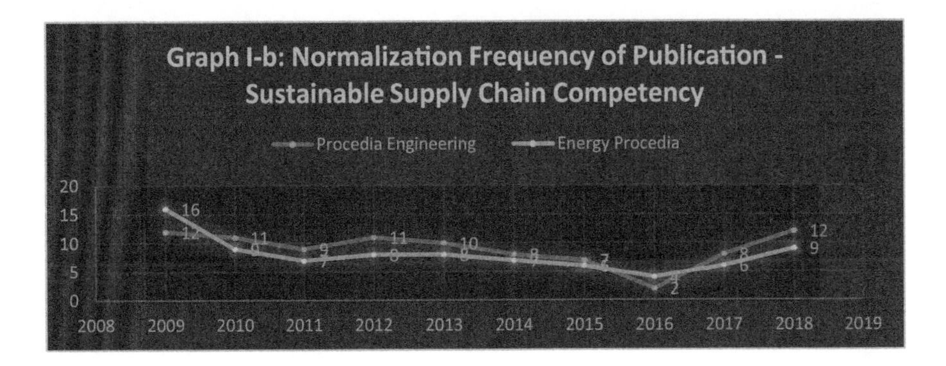

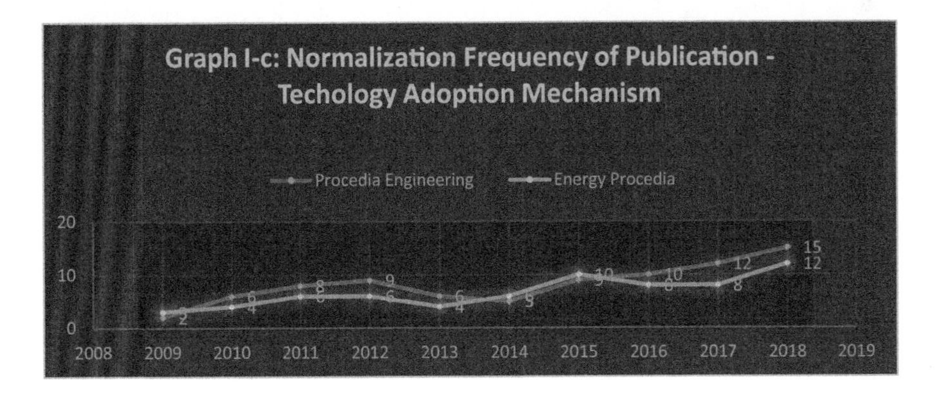

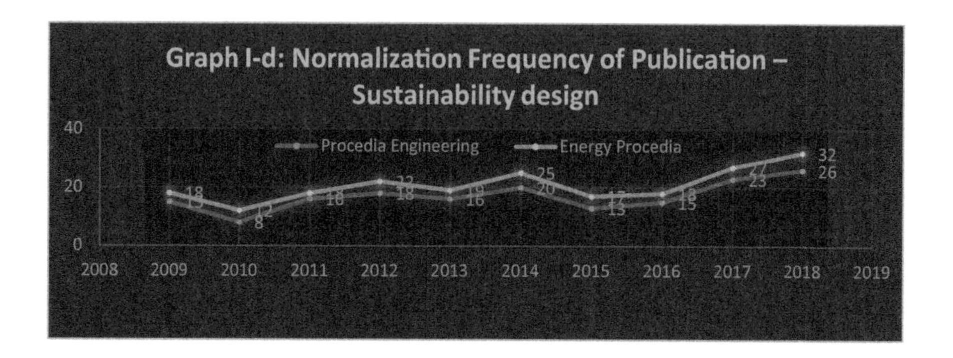
Graph I-d: Normalization Frequency of Publication – Sustainability design
Procedia Engineering
Energy Procedia
40
20
0
2008
2009
2010
2011
2012
2013
2014
2015
2016
2017
2018
2019
32
26

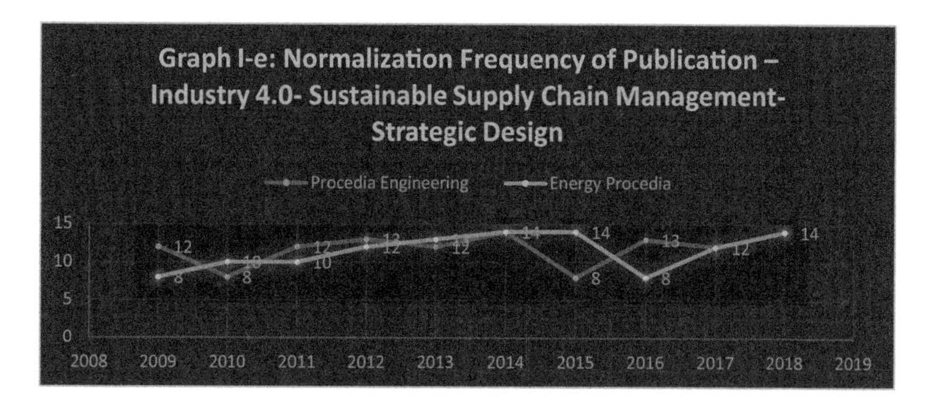
Graph I-e: Normalization Frequency of Publication – Industry 4.0- Sustainable Supply Chain Management- Strategic Design
Procedia Engineering
Energy Procedia
15
10
5
0
2008
2009
2010
2011
2012
2013
2014
2015
2016
2017
2018
2019

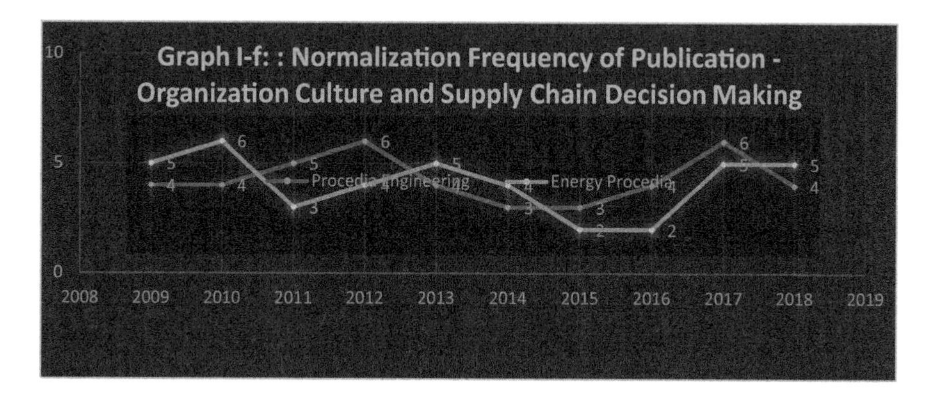
Graph I-f: : Normalization Frequency of Publication - Organization Culture and Supply Chain Decision Making
Procedia Engineering
Energy Procedia
10
5
0
2008
2009
2010
2011
2012
2013
2014
2015
2016
2017
2018
2019

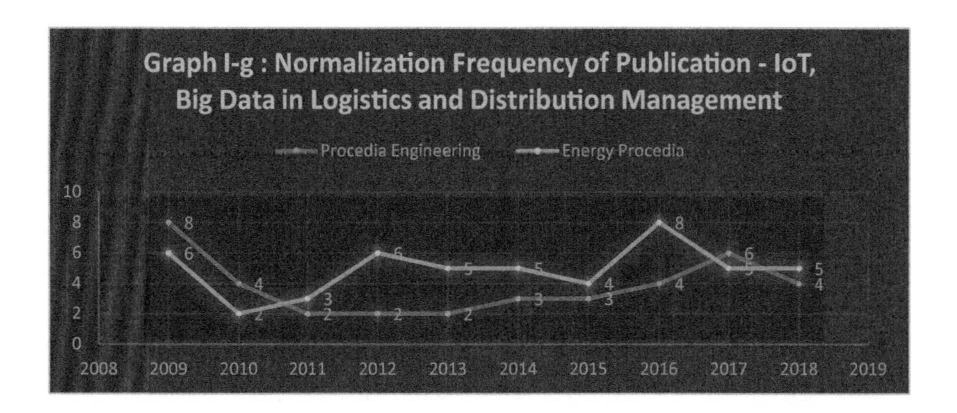

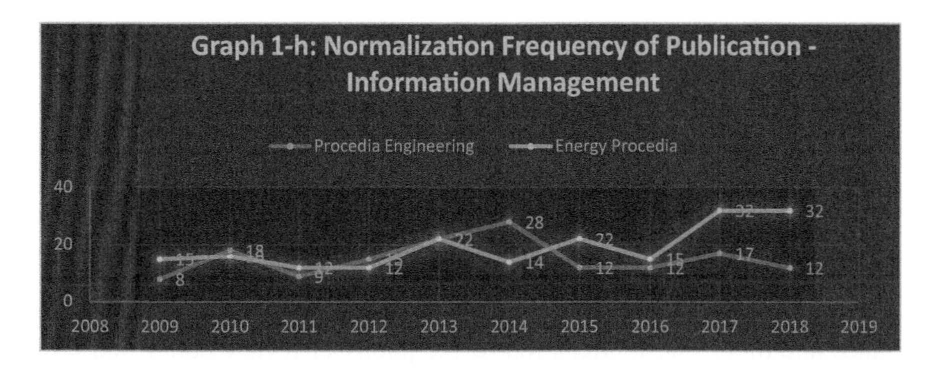

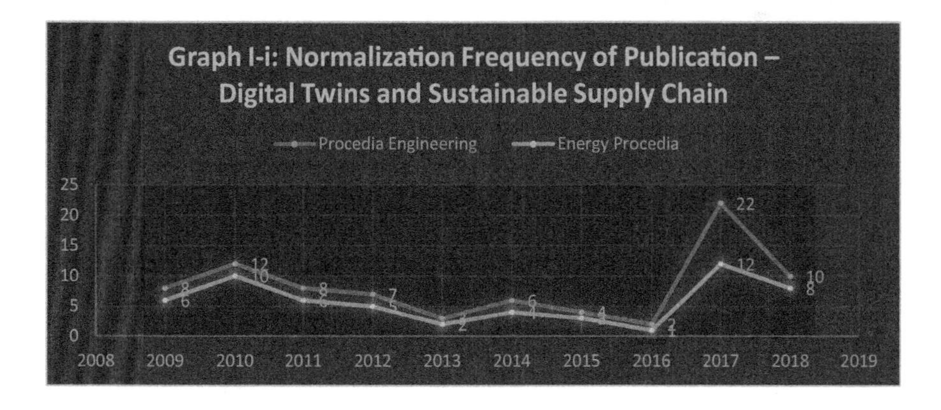

It has been observed that across the latent factors. There frequency can be depicted through the Chart 2.

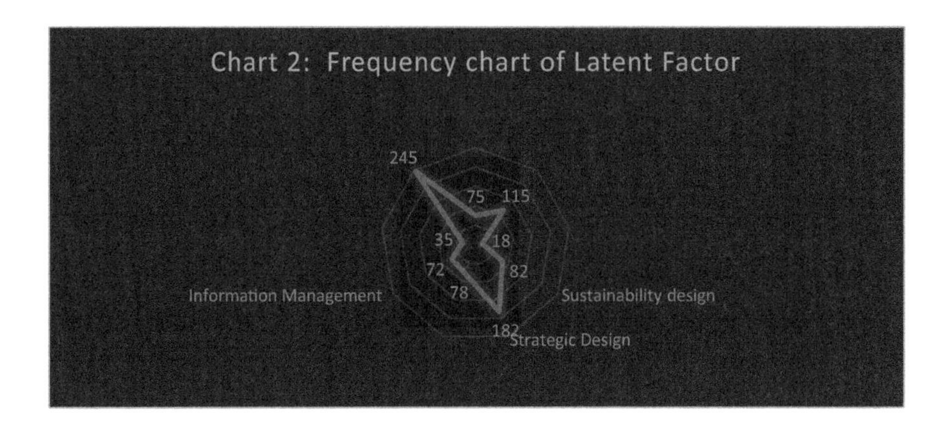

Chart 2: Frequency chart of Latent Factor

6.7 DISCUSSION

Top two well-known Open Access Conference Proceedings publication journals viz. Procedia Engineering and Energy Procedia are identified for analysis for the present study. These journals incorporates 82% of highest citations in the area of Industry 4.0 and Manufacturing and Service Operations and Supply Chain Management are identified. In total, both journals receives more than 2,546 submissions per year* and has the average acceptance rate of 17.5%. Journals follow the publisher review policy and take multiple rounds of major revisions to get a paper in an acceptable form. The number of reviewers and reviews per accepted paper are 2.4 and 4.6, respectively. Moreover, almost always, the reviewing process leading to acceptance goes through at least one major revision (Słowiński, 2016). The journals had average publication rate of 36 volumes per year and cover prestigious conferences in the area of Engineering Management, Technology, Sustainability, and Green Innovation. The publisher is Elsevier, Netherlands. Journals are indexed in the Journal Citation Reports (JCR) of the Web of Science (WoS) Core Collection database and Scopus. The search combination includes either of the Joint Operation. Viz.

1. "Smart Factory" and "Sustainability" and "Supply Chain" and "Service Operations"
2. "Service Operations" and "Sustainable development" and Technology-Based Supply Chains"
3. "Business" and "Supply Chain"
4. "Economics" and "Supply Chain"

5. "Industry 4.0" and "Cyber-Physical" and "Digital Twins"
6. "Sustainability" and "Service Supply Chain" and "Digitization"

In 2017, both journals, individually published 4,492 (Energy Procedia, Vol. 105–141) and 5,547 (Procedia Engineering, Vol. 170–210) articles, respectively. This paper presents a general bibliometric overview of the journal between 2009 (Vol. 1) and 2018 (present year). We identify and visualize leading trends in emerging Technologies including Smart Factory, IoT, Big Data, Cloud Computing, Digital – Twins, Artificial Intelligence (AI), Industry 4.0 and Supply Chain Management gained the moment since last 3 years.

Volume	Range	Number of Articles	(1)	(2)	(3)	(4)	(5)	(6)
Vol. 141	687	123	9	23	44	28	13	16
Vol. 140	505	45	3	8	16	10	5	6
Vol. 139	823	100	7	19	36	23	11	13
Vol. 138	1201	300	21	56	108	69	32	39
Vol. 137	577	57	4	11	21	13	6	7
Vol. 136	519	81	6	15	29	19	9	11
Vol. 135	521	48	3	9	17	11	5	6
Vol. 134	903	92	6	17	33	21	10	12
Vol. 133	443	39	3	7	14	9	4	5
Vol. 132	1023	171	12	32	62	39	18	22
Vol. 131	447	64	4	12	23	15	7	8
Vol. 130	145	22	2	4	8	5	2	3
Vol. 129	1155	149	10	28	54	34	16	19
Vol. 128	563	84	6	16	30	19	9	11
Vol. 127	431	50	4	9	18	12	5	7
Vol. 126	1161	147	10	27	53	34	15	19
Vol. 125	677	81	6	15	29	19	9	11
Vol. 124	951	124	9	23	45	29	13	16
Vol. 123	401	52	4	10	19	12	5	7
Vol. 122	1151	193	14	36	69	44	20	25
Vol. 121	307	40	3	7	14	9	4	5
Vol. 120	727	86	6	16	31	20	9	11

Volume	Range	Number of Articles	(1)	(2)	(3)	(4)	(5)	(6)
Vol. 119	1011	213	15	39	77	49	22	28
Vol. 118	251	69	5	13	25	16	7	9
Vol. 117	1189	238	17	44	86	55	25	31
Vol. 116	525	78	5	14	28	18	8	10
Vol. 115	513	87	6	16	31	20	9	11
Vol. 114	7665	456	32	84	164	105	48	59
Vol. 113	507	123	9	23	44	28	13	16
Vol. 112	689	167	12	31	60	38	18	22
Vol. 111	1079	349	24	65	126	80	37	45
Vol. 110	617	48	3	9	17	11	5	6
Vol. 109	495	42	3	8	15	10	4	5
Vol. 108	17	8	1	1	3	2	1	1
Vol. 107	397	31	2	6	11	7	3	4
Vol. 106	0	0	0	0	0	0	0	0
Vol. 105	5179	342	24	63	123	79	36	44

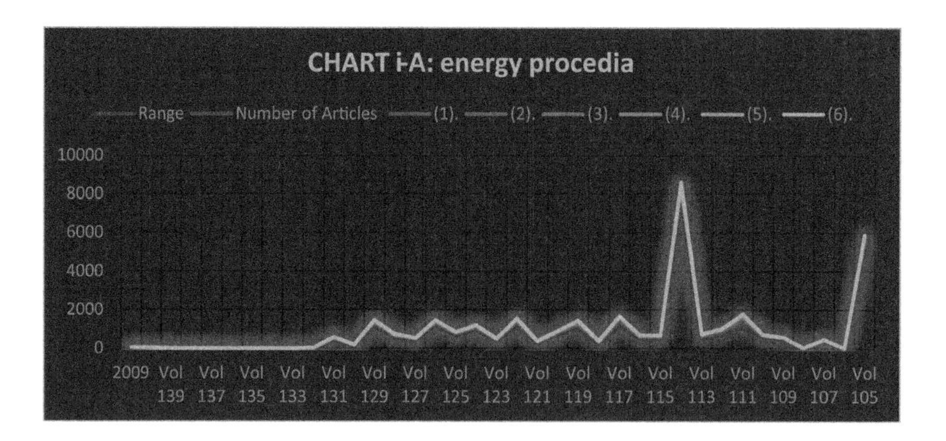

Volume	Range	Number of Articles	(1)	(2)	(3)	(4)	(5)	(6)
Vol. 170	557	543	76	33	14	35	24	14
Vol. 171	1549	377	53	23	9	25	17	9
Vol. 172	1309	76	11	5	2	5	3	2
Vol. 173	1999	65	9	4	2	4	3	2

Volume	Range	Number of Articles	(1)	(2)	(3)	(4)	(5)	(6)
Vol. 174	1409	765	107	46	19	50	34	19
Vol. 175	371	47	7	3	1	3	2	1
Vol. 176	739	122	17	7	3	8	5	3
Vol. 177	71	362	51	22	9	24	16	9
Vol. 178	613	34	5	2	1	2	2	1
Vol. 179	71	54	8	3	1	4	2	1
Vol. 180	1783	24	3	1	1	2	1	1
Vol. 181	1079	543	76	33	14	35	24	14
Vol. 182	785	35	5	2	1	2	2	1
Vol. 183	379	22	3	1	1	1	1	1
Vol. 184	783	24	3	1	1	2	1	1
Vol. 185	437	765	107	46	19	50	34	19
Vol. 186	683	47	7	3	1	3	2	1
Vol. 187	807	122	17	7	3	8	5	3
Vol. 188	507	362	51	22	9	24	16	9
Vol. 189	937	34	5	2	1	2	2	1
Vol. 190	689	54	8	3	1	4	2	1
Vol. 191	1217	544	76	33	14	35	24	14
Vol. 192	1025	543	76	33	14	35	24	14
Vol. 193	531	35	5	2	1	2	2	1
Vol. 194	551	22	3	1	1	1	1	1
Vol. 195	263	24	3	1	1	2	1	1
Vol. 196	1137	765	107	46	19	50	34	19
Vol. 197	303	47	7	3	1	3	2	1
Vol. 198	1147	122	17	7	3	8	5	3
Vol. 199	3587	362	51	22	9	24	16	9
Vol. 200	493	34	5	2	1	2	2	1
Vol. 201	861	54	8	3	1	4	2	1
Vol. 202	331	24	3	1	1	2	1	1
Vol. 203	425	32	4	2	1	2	1	1
Vol. 204	521	321	45	19	8	21	14	8
Vol. 205	4201	1282	179	77	32	83	58	32
Vol. 206	1865	897	126	54	22	58	40	22

Volume	Range	Number of Articles	(1)	(2)	(3)	(4)	(5)	(6)
Vol. 207	2409	289	40	17	7	19	13	7
Vol. 208	209	32	4	2	1	2	1	1
Vol. 209	223	42	6	3	1	3	2	1
Vol. 210	629	82	11	5	2	5	4	2

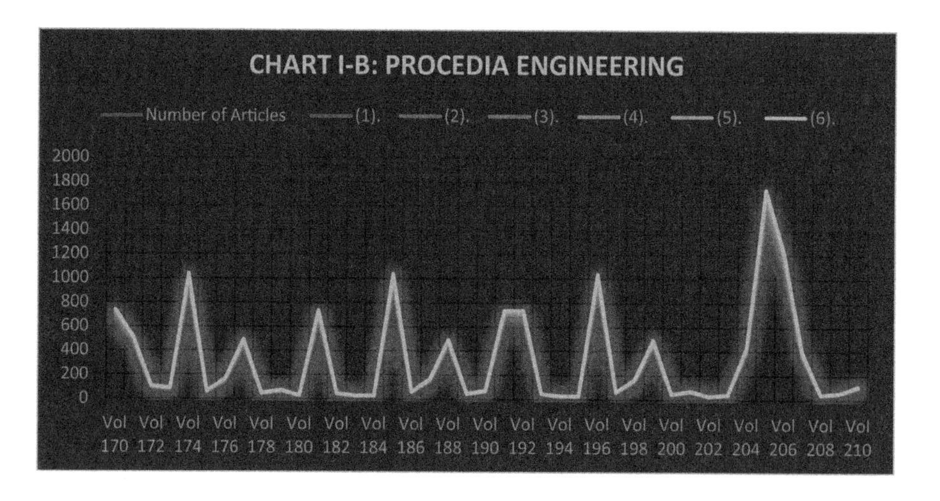

6.8 CONCLUSION

The chapter showcases the comprehensive state-of-art literature in the subject area of Industry 4.0 and Sustainable Supply Chain Management, with the aim to explore the growth of literature on Industry 4.0 across Manufacturing and Service Supply Chains. Research evidences are collected from noted list of 3561 articles published during 1991–2018 are taken into consideration belong to various databases (viz. Web of Science and Scopus, that includes, Social Citation Index Expanded (SCI-Expanded), Social Science Citation Index (SSCI), Art and Humanities Citation Index (A&HCI), conference proceedings, etc.).

LSA techniques were used as content analysis tool that combines text analysis and mining method to formulate latent factors. LSA creates the scientific ground to understand the trends. Using LSA, understanding future research trends will assist researchers in the area of Industry 4.0 and its applications in Manufacturing and Service Operations and Supply Chain

Management. The study will be beneficial for practitioners for strategic and operational aspect of service supply chain decision making.

KEYWORDS

- **Art and Humanities Citation Index**
- **industry 4.0**
- **Social Citation Index Expanded**
- **Social Science Citation Index**
- **sustainable supply chain management**

REFERENCES

Ahrens, D., & Spöttl, G., (2015). Industrie 4.0 und herausforderungen für die qualifizierung von fachkräften. In: *Digitalisierung industrieller Arbeit* (pp. 184–205). Nomos Verlagsgesellschaft mbH & Co. KG.

Baker, F. B., (1962). Information retrieval based upon latent class analysis. *Journal of the ACM (JACM)*, *9*(4), 512–521.

Boon-Itt, S., & Pongpanarat, C., (2011). Measuring service supply chain management processes: The application of the Q-sort technique. *International Journal of Innovation, Management, and Technology*, *2*(3), 217.

Brady, M. K., & Cronin, Jr., J. J., (2001). Some new thoughts on conceptualizing perceived service quality: A hierarchical approach. *Journal of Marketing*, *65*(3), 34–49.

Cho, D. W., Lee, Y. H., Ahn, S. H., & Hwang, M. K., (2012). A framework for measuring the performance of service supply chain management. *Computers & Industrial Engineering*, *62*(3), 801–818.

Chopra, S., & Meindl, P., (2007). Supply chain management. Strategy, Planning, and Operation. *Das Summa Summarum des Management*, 265–275.

Deerwester, S., Dumais, S. T., Furnas, G. W., Landauer, T. K., & Harshman, R., (1990). Indexing by latent semantic analysis. *Journal of the American Society for Information Science*, *41*(6), 391.

Evangelopoulos, N., Zhang, X., & Prybutok, V. R., (2012). Latent semantic analysis: Five methodological recommendations. *European Journal of Information Systems*, *21*(1), 70–86.

Gould, S. J., (1989). Punctuated equilibrium in fact and theory. *Journal of Social and Biological Structures*, *12*(2–3), 117–136.

Joshi, S., & Sarkis, J., (Forthcoming). Green collaborations, sustainability, and innovation in the emerging market economies: Multi-stakeholder perspective. In: *Greening of Industry*

4.0: Sustainable Supply Chain Value Creation through Transformation (pp. XX-XXX). Springer Netherlands.
Joshi, S., & Sharma, M., (2014). Blending green with lean-incorporating best-of-the-breed practices to formulate an optimum global supply chain management framework: Issues and concerns. In: *Handbook of Research on Design and Management of Lean Production Systems* (pp. 389–407). IGI Global.
Joshi, S., Sharma, M., & Rathi, S. (2017). Forecasting in Service Supply Chain Systems: A State-of-the-Art Review Using Latent Semantic Analysis. In Advances in Business and Management Forecasting (pp. 181-212). Emerald Publishing Limited.
Joshi, S., (forthcoming). Internet of things, big data and industry 4.0: Transformation and resilience strategies in supply chain and operations. *International Journal of Supply Chains and Operations Resilience, 4*(1).
Kagermann, H., Lukas, W. D., & Wahlster, W., (2015). Abschotten ist keine Alternative. VDI Nachrichten. *Technik–Wirtschaft–Gesellschaft, 16.*
Keating, B. A., Carberry, P. S., Bindraban, P. S., Asseng, S., Meinke, H., & Dixon, J., (2010). Eco-efficient agriculture: Concepts, challenges, and opportunities. *Crop Science, 50*(Supplement_1), S–109.
Keating, B., (2009). Managing ethics in the tourism supply chain: The case of Chinese travel to Australia. *International Journal of Tourism Research, 11*(4), 403–408.
Kotler, P., Kartajaya, H., & Setiawan, I., (2010). *Marketing 3.0: From Products to Customers to the Human Spirit.* John Wiley & Sons.
Kulkarni, S. S., Apte, U. M., & Evangelopoulos, N. E., (2014). The use of latent semantic analysis in operations management research. *Decision Sciences, 45*(5), 971–994.
Kundu, A., Jain, V., Kumar, S., & Chandra, C., (2015). A journey from normative to behavioral operations in supply chain management: A review using Latent Semantic Analysis. *Expert Systems with Applications, 42*(2), 796–809.
Kwantes, P. J., Derbentseva, N., Lam, Q., Vartanian, O., & Marmurek, H. H., (2016). Assessing the big five personality traits with latent semantic analysis. *Personality and Individual Differences, 102*, 229–233.
Lambert, D. M., Cooper, M. C., & Pagh, J. D., (1998). Supply chain management: Implementation issues and research opportunities. *The International Journal of Logistics Management, 9*(2), 1–20.
Landauer, T. K., (2007). LSA as a theory of meaning. *Handbook of Latent Semantic Analysis*, 3–34.
Levitt, T., (1960). Marketing Myopiaa. *Harvard Business Review, 38*(4), 24–47.
Lilley, O. L., (1954). Evaluation of the subject catalog. Criticisms and a proposal. *Journal of the Association for Information Science and Technology, 5*(2), 41–60.
Lisboa, P. J., & Taktak, A. F., (2006). The use of artificial neural networks in decision support in cancer: a systematic review. *Neural Networks, 19*(4), 408–415.
López, L., & Zúñiga, R., (2014). Dynamics of judicial service supply chains. *Journal of Business Research, 67*(7), 1447–1454.
Manhart, K., (2015). Potenzial für den Mittelstand. *Industrie 4.0-Die Nächste Revolution.*
Nouvellet, P., Cori, A., Garske, T., Blake, I. M., Dorigatti, I., Hinsley, W., ... & Fraser, C. (2018). A simple approach to measure transmissibility and forecast incidence. *Epidemics, 22*, 29–35.
Omar, Y. M., Minoufekr, M., & Plapper, P. (2018). Lessons from social network analysis to Industry 4.0. *Manufacturing Letters, 15*, 97–100.

Oztemel, E., & Gursev, S. (2018). Literature review of Industry 4.0 and related technologies. Journal of Intelligent Manufacturing, 1–56.

Paternò, F., & Santoro, C., (2002). Preventing user errors by systematic analysis of deviations from the system task model. *International Journal of Human-Computer Studies*, *56*(2), 225–245.

Rennung, F., Luminosu, C. T., & Draghici, A., (2016). Service provision in the framework of industry 4.0. *Procedia-Social and Behavioral Sciences*, *221*, 372–377.

Robertson, S. E., Van Rijsbergen, C. J., & Porter, M. F., (1980). Probabilistic models of indexing and searching. In: *Proceedings of the 3rd Annual ACM Conference on Research and Development in Information Retrieval* (pp. 35–56). Butterworth & Co.

Robertson, S., (2004). Understanding inverse document frequency: On theoretical arguments for IDF. *Journal of Documentation*, *60*(5), 503–520.

Salton, G., Wong, A., & Yang, C. S., (1975). A vector space model for automatic indexing. *Communications of the ACM*, *18*(11), 613–620.

Salton, G., Yang, C. S., & Yu, C. T., (1975). A theory of term importance in automatic text analysis. *Journal of the American Society for Information Science*, *26*(1), 33–44.

Santos, K., Loures, E., Piechnicki, F., & Canciglieri, O., (2017). Opportunities assessment of product development process in industry 4.0. *Procedia Manufacturing*, *11*, 1358–1365.

Seuring, S., & Gold, S., (2012). Conducting content-analysis based literature reviews in supply chain management. *Supply Chain Management: An International Journal*, *17*(5), 544–555.

Sidorova, A., Evangelopoulos, N., Valacich, J. S., & Ramakrishnan, T., (2008). Uncovering the intellectual core of the information systems discipline. *Mis Quarterly*, 467–482.

Siebel, T. M., (2017). Why digital transformation is now on the CEO's shoulders. *McKinsey Quarterly*, *12*, 12–26.

Siniscalchi, S. M., Svendsen, T., & Lee, C. H., (2014). An artificial neural network approach to automatic speech processing. *Neurocomputing*, *140*, 326–338.

Stock, T., & Seliger, G., (2016). Opportunities of sustainable manufacturing in industry 4.0. *ProcediaCorpp*, *40*, 536–541.

Sung, T. K., (2017). Industry 4.0: A Korea perspective. *Technological Forecasting and Social Change*.

Syntetos, A. A., Babai, Z., Boylan, J. E., Kolassa, S., & Nikolopoulos, K., (2016). Supply chain forecasting: Theory, practice, their gap and the future. *European Journal of Operational Research*, *252*(1), 1–26.

Tang, J., & Zhang, X., (2007). Quality-of-service driven power and rate adaptation over wireless links. *IEEE Transactions on Wireless Communications*, *6*(8).

Tapper, R., & Font, X., (2004). Tourism supply chains. Report of a desk research project for the travel foundation. *Tourism Supply Chains*.

Tarr, D., & Borko, H., (1974). Factors influencing inter-indexer consistency. In: *Proceedings of the ASIS 37th Annual Meeting* (Vol. 11, No. 974, pp. 50–55).

Thorleuchter, D., Scheja, T., & Van den Poel, D., (2014). Semantic weak signal tracing. *Expert Systems with Applications*, *41*(11), 5009–5016.

Van Der Zwan, F., & Bhamra, T., (2003). Services marketing: Taking up the sustainable development challenge. *Journal of Services Marketing*, *17*(4), 341–356.

Van Rijsbergen, C. J., (1977). A theoretical basis for the use of co-occurrence data in information retrieval. *Journal of Documentation*, *33*(2), 106–119.

Wang, K. T., (2015). *Research Design in Counseling*. Nelson Education.

Wiemer-Hastings, P., Wiemer-Hastings, K., & Graesser, A., (1999). Improving an intelligent tutor's comprehension of students with Latent Semantic Analysis. In: *Artificial Intelligence in Education* (Vol. 99).

Xu, L. D., Xu, E. L., & Li, L. (2018). Industry 4.0: state of the art and future trends. *International Journal of Production Research*, *56*(8), 2941–2962.

Yalcinkaya, M., & Singh, V., (2015). Patterns and trends in building information modeling (BIM) research: A latent semantic analysis. *Automation in Construction*, *59*, 68–80.

Yang, Z., & Yang, Y., (2006). Research and development of self-organizing maps algorithm. *Computer Engineering*, *16*(8), 201–203.

Zezulka, F., Marcon, P., Vesely, I., & Sajdl, O., (2016). Industry 4.0: an introduction in the phenomenon. *IFAC-Papers Online*, *49*(25), 8–12.

Zhong, R. Y., Xu, X., Klotz, E., & Newman, S. T., (2017). Intelligent manufacturing in the context of industry 4.0: A review. *Engineering*, *3*(5), 616–630.

CHAPTER 7

MULTIPLAYER COMPETITIVE MODEL GAMES AND ECONOMICS ANALYTICS

CYRUS F. NOURANI[1] and OLIVER SCHULTE[2]

[1]*SFU, Computation Logic Labs, Burnaby, BC, Canada; TU Berlin, AI, Germany, E-mail: cyrusfn@alum.mit.edu*

[2]*SFU, Computation Logic Labs, Burnaby, BC, Canada, E-mail: OSchulte@cs.sfu.ca*

ABSTRACT

A novel agent competitive learning model-computing techniques for analytics economics is developed with description computing logic models. Recalling a basis for model discovery and prediction planning from the first authors preceding years, new multiplayer game models are developed. The development is a basis for planning for predictive analytic support systems design. The techniques are forwarding a descriptive game logic where model compatibility is characterized on von Neumann, Morgenstern, Kuhn (VMK) game descriptions model embedding and game goal satisfiability. The techniques apply to both zero-sum and arbitrary games. The new encoding, with a *VMK game function situation,* where agent sequence actions, are have embedded measures. The import is that game tree modeling on VMK encodings are based on computable partition functions on generic game model diagrams. Newer payoff criteria on game trees and game topologies are obtained. Epistemic accessibility is addressed on game model diagrams payoff computations. Furthermore, criteria are presented to reach a sound and complete game logic based on VMK and hints on applications to compute Nash equilibrium criteria and a precise mathematical basis is stated.

7.1 INTRODUCTION

Modeling, objectives, and planning issues are examined with agent planning and competitive models. Model discovery and prediction are applied to compare models and get specific confidence intervals to supply to goal formulas. Competitive model learning is presented based on the new agent computing theories the first author had defined since 1994. The foundations are applied to present precise decision strategies on multiplayer games with only perfect information between agent pairs. The game tree model is applied to train models. The computing model is based on a novel competitive learning with agent multiplayer game tree planning. Specific agents transform the models to reach goal plans where goals are satisfied based on competitive game tree learning. Intelligent AND/OR trees and means-end analysis were applied with agents as the hidden –step computations.

A novel multiplayer game model is presented where "intelligent" agent enriched languages can be applied to address game questions on models in the mathematical logic sense. The game tree model is applied to encode von Neumann, Morgenstern, Kuhn games that were rendering on a situation calculus on (Schulte and Delgrande, 2004). The sections outline is as follows. Section 7.2 briefs on knowledge representation on game trees and games as plans.

Minimal predictions and predictive diagrams are examined on applications to game trees as goal satisfaction. Section 7.3 briefs on competitive models as a basis for realizing agent game trees as goals. Agent AND/OR trees are presented to carry on game tree computations. Infinitary multiplayer games are examined and modeling questions with generic diagrams are studied. Basic two person agent games and the group characterizations based on the first authors preceding publications are briefed to lead onto a newer area the second author and company had developed on von Neumann, Morgenstern, Kuhn games. Games, situations, and compatibility are stated with a new encoding on the above game situations that are based on game tuple ordinals computed on generic diagrams. Game descriptions based on the first author's description logic on generic diagrams are applied to KVM to reach new model-theoretic game tree bases. The new KVM encoding, *VMK game function situation, and game tree model diagrams* embed aggregate measures as lifted preorders based on the game tree nodes information partition functions allows us to embed measures on game trees where agent sequence actions, be that random sequences are modeled. The import is that game tree modeling, KVM encodings based on partition functions on generic game model diagrams, allows us to reach newer computability areas on game trees based on agent game trees on KVM game models. Novel payoff criteria on game trees and game topologies are obtained. Applying that to a game tree diagram models and generic diagram embeddings, payoff function characterizations are presented based on game tree topologies. Epistemic accessibility is addressed on game model diagrams payoff computations. Section 7.5 presents criteria on a sound and complete game logic based on KVM and hints on applications to compute Nash equilibrium criteria. Furthermore, games and tractability areas are addressed based on descriptive computing, cardinality on concept descriptions, and descriptive computability. Section 7.6 briefs on computing games on proof trees based on plan goal satisfaction, predictions, and game tree models, following the first author's publications past decade.

7.2 KR PLAN DISCOVERY

Modeling with agent planning is applied where uncertainty, including effect or and sensor uncertainty, are relegated to agents, where competitive learning on game trees determines a confidence interval. The incomplete knowledge modeling is treated with KR on predictive model diagrams. Model discovery at KB's is with specific techniques defined for trees. Model diagrams allow us to model-theoretically characterize incomplete KR. To key into the incomplete knowledge base, we apply generalized predictive diagrams whereby specified diagram functions a search engine can select onto localized data fields. The predictive model diagrams (Nourani, 1995) could be minimally represented by the set of functions {f1,...,fn}. Models uphold to a deductive closure of the axioms modeled and some rules of inference. A generic model diagram (G-diagram) (Nourani, 1991, 1994a) is a diagram in which the elements of the structure are all represented by a specified minimal set of function symbols and constants, such that it is sufficient to define the truth of formulas only for the terms generated by the minimal set of functions and constant symbols. Such assignment implicitly defines the diagram. It allows us to define a canonical model of a theory in terms of a minimal family of function symbols. The minimal set of functions that define a G-diagram are those with which a standard model could be defined.

7.2.1 PREDICTIONS AND DISCOVERY

A *predictive diagram* for a theory T is a diagram D (M), where M is a model for T, and for any formula q in M, either the function f: q → {0,1} is defined, or there exists a formula p in D (M), such that T{p} proves q; or that T proves q by minimal prediction. A *generalized predictive diagram* is a predictive diagram with D (M) defined from a minimal set of functions. The predictive diagram could be minimally represented by a set of functions {f1,...,fn} that inductively define the model. The free trees we had defined by the notion of probability implied by the definition, could consist of some extra Skolem functions {g1,...,gm} that appear at free trees. The *f* terms and *g* terms, tree congruences, and predictive diagrams then characterize partial deduction with free trees. The predictive diagrams are applied to discover models to the intelligent game trees. Prediction is applied to plan goal satisfiability and can be combined with plausibility (Nourani, 1991) probabilities, and fuzzy logic to obtain; for example, confidence intervals.

7.2.2 BUSINESS MODELING EXAMPLES

The following examples from Nourani (2010) can be motivating for business applications. Example: A business manager has 6 multitalented players, designed with personality codes indicated with codes on the following balls. The plan is to accomplish five tasks with persons with matching personality codes to the task, constituting a team. Team competitive models can be generated by comparing teams on specific assignments based on the task area strength. The optimality principles outlined in the first author's publications might be to accomplish the goal with as few a team grouping as possible, thereby minimizing costs (Figure 7.1).

The following section presents new agent game trees the author had put forth (Nash, 1951). Applying game theory to business is tantamount to an interactive decision theory. Decisions are based on the world as given. However, the best decision depends on what others do, and what others do may depend on what they think you do. Hence games and decisions are intertwined. A second stage business plan needs to specify how to assign resources with respect to the decisions, ERP plans, and apply that to elect supply chain policies, which can in part specify how the business is to operate. A tactical planning model that plans critical resources up to sales and delivery is a business planner's dream. Planning and tasking require

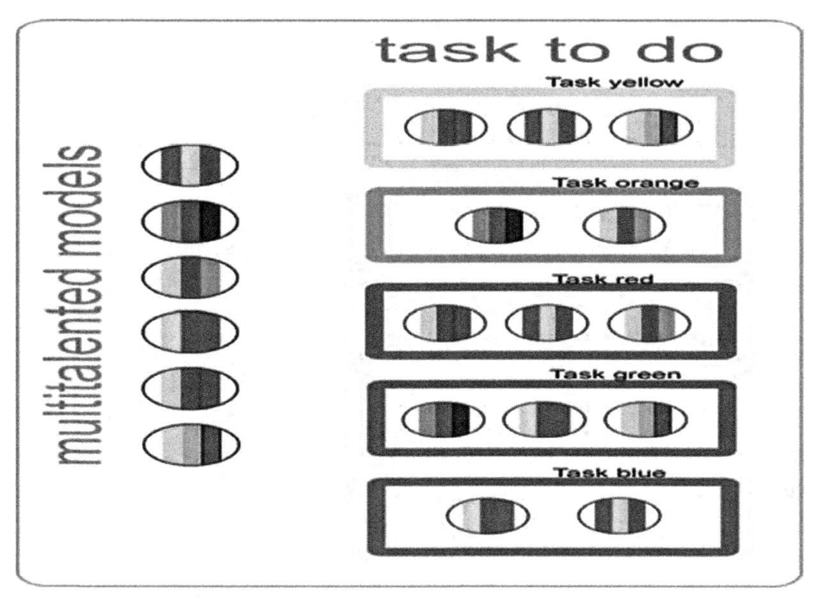

FIGURE 7.1 (See color insert.) Multitalented models.

the definition of their respective policies and processes; and the analyses of supply chain parameters. The above are the key elements of a game, anticipating behavior and acquiring an advantage. The players on the business planned must know their options, the incentives, and how do the competitors think.

Example premises: Strategic interactions.

Strategies: {Advertise, Do Not Advertise} Payoffs: Companies' Profits Advertising costs USD million.

The branches compute a Boolean function via agents. The Boolean function is what might satisfy a goal formula on the tree. An intelligent AND/OR tree is solved if the corresponding Boolean functions solve the AND/OR trees named by intelligent functions on the trees. Thus, node m might be *f* (a1,a2,a3) & *g* (b1,b2), where *f* and *g* are Boolean functions of three and two variables, respectively, and ai's and bi's are Boolean-valued agents satisfying goal formulas for *f* and g. The agent trees are applied to satisfy goals to attain competitive models for business plans and ERP.

The AND vs. OR principle is carried out on the above trees with the System to design ERP systems and manage as Cause principle decisions.

The agent business modeling techniques the author had introduced (Koller & Pfeffer, 1997; Schulte & Delgrande, 2004) apply the exact 'system as cause' and 'system as symptom' based on models (assumptions, values, etc.) and the 'system vs. symptom' principle via tracking systems behavior with cooperating computational agent trees. The design might apply agents splitting trees, where splitting trees is a well-known decision tree technique. Surrogate agents are applied to splitting trees. The technique is based on the first author's intelligent tree project (ECAI 1994 and European AI Communication) journal are based on agent splitting tree decisions like what is designed later on the CART system: The ordinary splitting tree decisions are regression-based, developed at Berkeley and Stanford (Brieman, Friedman et al., 1984; Breiman, 1996). The agent splitting agent decision trees have been developed independently by the author since 1994 at the Intelligent Tree Computing project. For new directions in forecasting and business planning (Nourani, 2002). Team coding example diagram from reach plan optimal games where a project is managed with a competitive optimality principle is to achieve the goals minimizing costs with the specific player code rule first author and company 2005.

7.3 COMPETITIVE MODELS GAMES

Planning is based on goal satisfaction at models. Multiagent planning is modeled as a competitive learning problem where the agents compete on game trees as candidates to satisfy goals hence realizing specific models where the plan goals are satisfied (for example, Muller-Pischel, 1994, Bazier et al., 1997). When a specific agent group "wins" to satisfy a goal the group has presented a model to the specific goal, presumably consistent with an intended world model. Therefore, plan goal selections and objectives are facilitated with competitive agent learning. The intelligent languages (Nourani, 1993d) are ways to encode plans with agents and compare models on goal satisfaction to examine and predict via model diagrams why one plan is better than another or how it could fail.

Games play an important role as a basis to economic theories. Here, the import is brought forth onto decision tree planning with agents. Intelligent tree computing theories we have defined since 1994 can be applied to present precise strategies and prove theorems on multiplayer games. Game tree degree with respect to models is defined and applied to prove soundness and completeness. The game is viewed as a multiplayer game with only perfect information between agent pairs. Upper bounds on determined games are presented. The author had presented a chess-playing basis in 1997 to a computing conference.

For each chess piece, a designating agent is defined. The player P makes its moves based on the board B it views. <P,B> might view chess as if the pieces on the board had come alive and were autonomous agents carrying out two-person games as in Alice in Wonderland. Game moves are individual tree operations.

7.3.1 INTELLIGENT AND/OR TREES AND SEARCH

AND/OR trees (Nilsson, 1969) are game trees defined to solve a game from a player's standpoint (Figure 7.2).

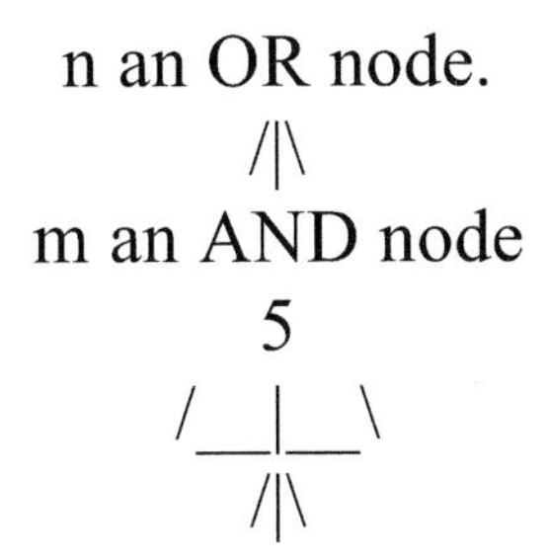

FIGURE 7.2 AND/OR trees.

Formally a node problem is considered solved if one of the following conditions hold. (1) The node is the set of terminal nodes (primitive problem-the node has no successor). (2) The node has AND nodes as successors and the successors are solved. (3) The node has OR nodes as successors and any one of the successors is solved. A solution to the original problem is given by the subgraph of AND/OR graph sufficient to show that the node is solved. A program which can play a theoretically perfect game would have a task like searching and AND/OR tree for a solution to a one-person problem to a two-person game. An intelligent AND/OR tree is where the tree branches are intelligent trees (Nourani, 1991). The branches compute a Boolean function via agents. The Boolean function is what might satisfy a goal formula on the tree. An intelligent AND/OR tree is solved if *f* the corresponding Boolean functions solve the AND/OR trees named by intelligent functions on the trees. Thus, node m might be *f*(a1,a2,a3) & *g*(b1,b2), where *f* and *g* are

Boolean functions of three and two variables, respectively, and ai's and bi's are Boolean-valued agents satisfying goal formulas for *f* and *g* (Figure 7.3).

```
g is on OR agent
      /|\
       |
   b1 | b2 f
    /__|__\
      /|\
f is an AND agent
    a1 a2 a3
```

FIGURE 7.3 Agent AND/OR trees.

The chess game trees can be defined by agent augmenting AND/OR trees (Nourani, (1990s). For the intelligent game trees and the problem-solving techniques defined, the same model can be applied to the game trees in the sense of two-person games and to the state space from the single agent view. The two-person game tree is obtained from the intelligent tree model, as is the state space tree for agents. To obtain the two-person game tree the cross-board-coboard agent computation is depicted on a tree. Whereas the state-space trees for each agent are determined by the computation sequence on its side of the board-coboard. A *game tree degree* is the game state a tree is at with respect to a model truth assignment, e.g., to the parameters to the Boolean functions above. Let generic diagram or G-diagrams be diagrams definable by specific functions. Intelligent signatures (Nourani, 1996a) are signatures with designated multiplayer game tree function symbols. A soundness and completeness theorem is proved on the intelligent signature language (Nourani, 1999a). The techniques allowed us to present a novel model-theoretic basis to game trees, and generally to the new intelligent game trees. The following specifics are from Nourani (1999). Let N be the set of all functions from w to w. Let A be a subset of N. (Gale-Stewart, 1953) associated with a 2-person game of perfect information G<A>. Player I begins by choosing n0 in w; player two chooses n1 in w; then I choose n2 in w; so on. Let a(i) = ni. I wins G<A> if and only if a in A. We say that G<A> is determined if one of the players has a winning strategy.

Proposition 1: If G<A> is determined, the complexity upper bound on the number of moves to win is A's cardinality.

Theorem 2: For every pair p of opposing agents there is a set A<p> (N. The worse case bound for the number of moves for a determined game based on the intelligent game tree model is the sum ({|A<p>|: p agent pairs}.

Proof: Sum over the proposition (Nourani, 1999). At the intelligent game trees, the winning agents determine the specific model where the plan goals are satisfied.

In the present chapter, we are considering game models, zero-sum, or not, however not carrying on with intent agent analytics (Nourani & AAAI, 2002). At the agent intelligent game trees, the winning agents determine the specific model where the plan goals are satisfied. On what follows we first examine game situations and models that can be applied to zero-sum or nondeterministic based on game goals are satisfied on game tree models.

7.3.2 GAMES, SITUATIONS, AND COMPATIBILITY

We begin with a basic mathematical definition of what situations are followed with what model diagrams are and we can characterize situations with generic model diagrams.

Definition 3.1: A situation consists of a nonempty set D, the domain of the situation, and two mappings: *g* and *h*. *g* is a mapping of function letters into functions over the domain as in standard model theory. h maps each predicate letter, pn, to a function from Dn to a subset of {t,f}, to determine the truth value of atomic formulas as defined below. The logic has four truth values: the set of subsets of {t,f}.{{t},{f},{t,f},0}. The latter two is corresponding to inconsistency, and lack of knowledge of whether it is true or false.

Due to the above truth values, the number of situations exceeds the number of possible worlds. The possible worlds being those situations with no missing information and no contradictions. From the above definitions, the mapping of terms and predicate models extend as in standard model theory. Next, a *compatible set of situations* is a set of situations with the same domain and the same mapping of function letters to functions. In other words, the situations in a compatible set of situations differ only on the truth conditions they assign to predicate letters. Thus, the diagram might be considered as a basis for a model, provided we can by algebraic extension, defines the truth value of arbitrary formulas instantiated with arbitrary terms. Thus, all compound sentences build out of atomic sentences then could be assigned a truth value, handing over a model. It will be made clearer in the following subsections. The following examples would run throughout the paper. Consider the primitive first-order language (FOL) from the first authors, a decade's publications encoding with generic world models specifics are as follows.

L = {c},{f(X)},{p(X),q(X)}

Let us apply Prolog notation convention for constants and variables) and the simple theory {for all X: p(X) → q(X),p(c)}, and indicate what is meant by the various notions.

[model] = {p(c),q(c),q(f(c)),q(f(f(c)))},{p(c) ∧ q(c), p(c) ∧ p(X), p(c) & p(f(X))}, {p(c) v p(X), p(c) v p(f(X)), p(c) → p(c)}

[diagram] = {p(c),q(c),p(c),q(f(c)),q(f(f(c)))},q(X)}; i.e., diagram = the set of atomic formulas of a model.

Thus, the diagram is [diagram] = {p(c),q(c),q(f(c)),q(f(f(c))),...,q(X)}

Based on the above, we can define generalized diagrams. The term generalized is applied to indicate that such diagrams are defined by algebraic extension from basic terms and constants of a language. The fully defined diagrams make use of only a minimal set of functions. Generalized diagram is [generalized diagram] = {p(c),q(c),p(f(t)),q(f(t))}, for t defined by induction, as {t_0 = c, and t_n = {f(t(n – 1))} for n > 0.

It is thus not necessary to redefine all f(X)'s since they are instantiated. Nondeterministic diagrams are those in which some formulas are assigned an indeterminate symbol, neither true nor false, that can be arbitrarily assigned in due time.

[nondeterministic diagram] = {p(c),q(c),p(f(t)),q(f(c)),q(f(f(c))),I_q(f(s)))}, where *t* is as defined by induction before and I_q(f(s)) = I_q for some indeterminate symbol I_q, for {s = t_n, n >= 2. These G-diagrams are applicable for KR in planning with incomplete knowledge with backtracking *Free Skolemized diagrams* are those in which instead of an indeterminate symbols there are free Skolem functions, that could be universally quantified.

[free Skolemized diagram] = {p(c),q(c),p(f(t)),q(f(c)),q(f(f(c))), q_F.s(s)}, where t and s are as defined in the sections before. These G-diagrams are applicable to KR for planning with incomplete knowledge (Nourani, 1991) and free proof trees.

A **generalized free diagram** (GF-diagram) is a diagram that is defined by algebraic extension form a minimal set of function symbols.

[generalized free diagram] = {p(c),q(c),p(f(t)),q(f(t)), q_F.s(s)}, for t and s as before. A generalized free plausible diagram is a GFD that has plausibility degrees assigned to it. These G-diagrams are applied for KR in planning with free proof trees (Nourani, 1991; Vulkan, 2012).

7.3.3 GAME TREE PROJECTIONS ON MODEL DIAGRAMS

Game tree diagrams are encoded with model diagram functions. Repeating the example from the preceding sections. A game tree node m might be f(a1,a2,a3) & g(b1,b2), where *f* and *g* are Boolean functions of three and two variables, respectively, and ai's and bi's are Boolean-valued agents satisfying goal. On the above game model diagram, the tree is defined with two terms f(a1,a2,a3) & g(b1,b2) on a two-stage collapse to a model diagram that can be assigned a Boolean value. Typical assignments might be in general: f(t1…,tn)…{T,F, H(t1…,tn)}, when t1…,tn are free Skolemized agent trees, and h is a tupled morphism <h1,…,hn> on the undetermined trees on a Boolean algebra, T and F, are True and False assignments.

Definition 3.2: Let M be a structure for a language L, call a subset X of M a generating set for M if no proper substructure of M contains X, i.e., if M is the closure of X U {c[M]: c is a constant symbol of L}. An assignment of constants to M is a pair <A,G>, where A is an infinite set of constant symbols in L and G: A → M, such that {G[a]: a in A} is a set of generators for M. Interpreting a by g[a], every element of M is denoted by at least one closed term of L[A]. For a fixed assignment <A,G> of constants to M, the diagram of M, D<A,G>[M] is the set of basic [atomic and negated atomic] sentences of L[A] true in M. [Note that L[A] is L enriched with set A of constant symbols.]

Definition 3.3: A Generic diagram for a structure M is a diagram D<A,G>, such that the *G* in definition 3.2 has a proper definition by specific function symbols.

The dynamics of epistemic states as formulated by generalized diagrams (Nourani, 1998, p.91) is exactly what addresses the compatibility of situations. What it takes to have an algebra and model theory of epistemic states, as defined by the generalized diagram of possible worlds is what (Nourani, 1998, p.91) carries on with. To decide compatibility of two situations we compare their generalized diagrams. Thus, we have the following theorem.

Theorem 3 (Nourani 1994): Two situations are compatible if their corresponding generalized diagrams are compatible with respect to the Boolean structure of the set to which formulas are mapped (by the function h above, defining situations).

Proof: The generic diagrams, definition 4.3, encode possible worlds and since we can define a one-to-one correspondence between possible worlds and truth sets for situations (Nourani, 1998, 1991, 1984), computability is definable by the G-diagrams.

The computational reducibility areas were briefed at Nourani (1998, 2009). An implication is that with infinitary logic we can form an infinite conjunction of beliefs with respect to possible worlds. The worlds are represented with the compatible generalized diagrams that can characterize reaching for an object form the model-theoretic point of view, for example in a Kantian sense, or reaching goals on game trees.

7.4 DESCRIPTIONS AND GAMES

Von Neumann and Morgenstern (VM for short) designed game theory as a very general tool for modeling agent interactions. Schulte et al. (2004) show that every such VM game *G* has an axiomatization Axioms(G) in the situation calculus that represents the game in the following strong sense: The game *G* itself is a model of Axioms(G), and all models of Axioms(G) are isomorphic. It follows that the axiomatization is correct, in the sense that Axioms (G) entails only true assertions about the game G, and complete in the sense that Axioms(G) entails all true assertions about the game G.

Let us call axiomatizations A(G) where all models of A(G) are isomorphic a categorical axiomatization.

Definition 4.4: (von Neumann, Morgenstern, Kuhn). A sequential game *G* is a tuple ⟨N,H,player,fc,{Ii},{ui}⟩ whose components are as follows.

1. A finite set N (the set of *players)*.
2. A set H of sequences satisfying the following three properties.
   (a) The empty sequence ∅ is a member of H.
   (b) If σ is in H, then every initial segment of σ is in H.
   (c) If h is an infinite sequence such that every finite initial segment of h is in H, then h is in H.

Each member of H is a *history;* each element of a history is an *action* taken by a player. A history σ is *terminal* if there is no history $s \in H$ such that $\sigma \subseteq s$. (Thus all infinite histories are terminal.) The set of terminal histories is denoted by Z. The set of actions available at a finite history s is denoted by $A(s) = \{a: s \subseteq a \in H\}$.
A function *player* that assigns to each nonterminal history a member of $N \cup \{c\}$. The function *player* determines which player takes an action after the history s. If a player(s) = c, then it is "nature's" turn to make a chance move.

3. A function fc that assigns to every history s for which player(s) = c a probability measure $fc(\cdot|s)$ on A(s). Each probability measure is independent of every other such measure. (Thus $fc(a|s)$ is the probability that "nature" chooses action a after the history s.)
4. For each player $i \in N$ an *information partition* Ii defined on $\{s \in H: \text{player}(s) = i\}$. An element Ii of Ii is called an *information set* of Player i. We require that if s, s′ are members of the same information set Ii, then $A(s) = A(s')$.
5. For each player, $i \in N$ a *payoff function* ui: $Z \to R$ that assigns a real number to each terminal history.

An element *Ii* of *Ii* is called an *information set* of Player *i.* We require that if *s, s* are members of the same information set *Ii,* then *A(s)* = *A(s).* 6. For each player, $i \in N$ a *payoff function ui:* $Z \to R$ that assigns a real number to each terminal history.

To examine game models we present an algebraic characterization that are not specific on the game so far as the probability measure assignment on game sequences are concerned. The information set on the game nodes are encoded on the game diagram. Whatever the game degree is at the node is how the following sequence is carried on. On that basis when can examine games based on satisfying a goal, where the compatibilist is characterized on specific VMK model embedding that is stated in the following section. Let us state a categorical VMK situation. For the time being, consider modeling zero-sum games and not adversarial games not to confuse the modeling questions. Modeling agent games based on game tree satisfiability does not require explicit probability measures. The information partition can be encoded such that whatever the unknown probabilities are, that in any case are very difficult to ascertain, the game tree node moves can be determined as the game develops. In Section 3.7, we present a treatment on measures and payoffs.

Definition 4.5: A ***VMK game function situation*** consists of a domain *N x H, where N is the natural numbers, H a set of game history sequences, and a mapping pair g,* f. *f maps function letters to (agent) functions and g maps pairs from <N,H> to {t.f}.*

Notes: On condition 5 at VMK definition 4.4 the information partition for a player I is determined at each game node by the game diagram game tree degree (Section 4.1) at that node that is known to that player, c.f. (Nourani, 1994, 1996).

The basic intricacy is what can you say about conditions on H on the encoding on a basic definition: H for example: that agent player plays to increases the pay on ordinals on every move. That is, there is an ordering on the functions on <N,H>, that is from a mathematical absolute. That is essentially a pre-ordered mapping to {t,f} where the ordering is completed on stages. However, the ordering is not known before a game goal is satisfied.

A. Models Preliminaries: To have precise statements on game models we begin with basic model-theoretic preliminaries. Agent models and morphisms on agent models had been treated in the first author's publications since the past decade. We develop that area further to reach specific game models. We apply basic model theory techniques to check upward compatibility on game tree goal satisfaction. The modeling techniques are basic game tree node stratification. Since VMK has a "categorical acclimatization" we can further explore game model embedding techniques with morphisms that the first author has developed on categorical models during the past decade and applied to agent computing (Nourani, 1996, 2005).

Let us start from certain model-theoretic premises with propositions known form basic model theory. A structure consists of a set along with a collection of finitary functions and relations, which are defined on it. Universal algebra studies structures that generalize the algebraic structures such as groups, rings, fields, vector spaces, and lattices. Model theory has a different scope than universal algebra, encompassing more arbitrary theories, including foundational structures such as models of set theory. From the model-theoretic point of view, structures are the objects used to define the semantics. of first-order logic. A structure can be defined as a triple consisting of a domain A, a signature Σ, and an interpretation function I that indicates how the signature is to be interpreted on the domain. To indicate that a structure has a particular signature σ one can refer to it as a σ-structure. The domain of a structure is an arbitrary set (non-empty); it is also called the underlying set of the structure, its carrier (especially in universal algebra), or its universe (especially in model theory). Sometimes the notation or is used for the domain of, but often no notational distinction is made between a structure and its domain, i.e., the same symbol can refer to both. The signature of a structure consists of a set of function symbols and relation symbols along with a function that ascribes to each symbol s a natural number, called the arity of s.

For example, let *G* be a graph consisting of two vertices connected by an edge, and let H be the graph consisting of the same vertices but no edges. H is a subgraph of G, but not an induced substructure. Given two structures and of the same signature Σ`, a Σ-homomorphism is a map that can preserve the functions and relations. A Σ-homomorphism h is called a Σ-embedding if it is one-to-one and for every n-ary relation symbol R of Σ and any elements ai..., an, the following equivalence holds: R(a1...,an) iff R(h(a1).h(an)).

Thus, an embedding is the same thing as a strong homomorphism, which is one-to-one. A structure defined for all formulas in the language consisting of the language of A together with a constant symbol for each element of M, which is interpreted as that element. A structure is said to be a model of a theory T if the language of M is the same as the language of T and every sentence in T is satisfied by M. Thus, for example, a "ring" is a structure for the language of rings that satisfies each of the ring axioms, and a model of ZFC set theory is a structure in the language of set theory that satisfies each of the ZFC axioms. Two structures M and N of the same signature σ are elementarily equivalent if every first-order sentence (formula without free variables) over σ is true in M if and only if it is true in N, i.e., if M and N have the same complete first-order theory. If M and N are elementarily equivalent, written M ≡ N.

A first-order theory is complete if and only if any two of its models are elementarily equivalent. N is an elementary substructure of M if N and M are structures of the same signature ∑ such that for all first-order ∑-formulas φ(x1...,xn) with free variables x1...,xn, and all elements a1...,an of N, φ(a1...,an) holds in N if and only if it holds in M: N |= φ(a1...,an) iff M |= φ(a1...,an). Let M be a structure of signature R(a1...,an), there are b1...,bn such that R(h(a1).,h(an)) and N a substructure of M. N is an elementary substructure of M if and only if for every first-order formula φ(x,y1...,yn) over σ and all elements b1...,bn from N, if M x\φ(x,b1...,bn), then there is an element a in N such that M φ(a,b1...,bn). An elementary embedding of a structure N into a structure M of the same signature is a map h: N → M such that for every first-order σ-formula φ(x1...,xn) and all elements *a1...,an* of *N, N* |= φ(a1...,an) implies *M* |= φ(h(a1).,h(an)). Every elementary embedding is a strong homomorphism, and its image is an elementary substructure.

Proposition 2: Let A and B be models for a language L. Then, A is isomorphically embedded in B iff B can be expanded to a model of the diagram of A

B. VMK and Axiom Games: From the preceding sections we remind ourselves that on VMK games when a game *G* is a model of *Axioms (G),* all models of *Axioms (G)* are isomorphic. *Axioms (G)* entails *only* true assertions about the game *G,* and *complete* in the sense that *Axioms (G)* entails *all* true assertions about the game G. Situation calculus is applied to model the structure of a multi-agent interaction, we consider agents *reasoning* about optimal actions in the multi-agent system. We show how to define the game-theoretic notion of a *strategy* or policy, and introduce predicates describing which strategies are optimal in a given environment and which strategy combinations form Nash equilibriums.

The present chapter we apply the generalized diagram encodings to situation calculus (Nourani, 1994–2010) on a descriptive computing to VMK, presenting a computable model-theoretic characterization. Descriptive definability on model diagrams was presented at Nourani (2000) and modal diagram considerations at Nourani (1998) on the Uppsala Logic Colloquium. To start with we are checking on VMK models where agent games can have variable function games assigned on the arbitrary basis at game tree nodes. The game becomes nondeterministic in that sense but we are not assigning probability measures at all. We set-up a premise that a player at a node takes game tree route that increases game tree degree, an ordinal, support. The VMK diagram models can carry that.

C. Compatibility Ordering and VMK Game Trees: Agent game tree generic diagram satisfiability based on the above compatibility criteria on game tree nodes implies the following.

Definition 4.6: Two generic diagrams D1 and D2 on the same logical language £ are compatible iff forever assignment to free variables to models on £, every model M for £, M $\models$ D1 iff M $\models$ D2.

Proposition 3: A game tree node generic diagram on a player at an agent game tree node is satisfiable iff the subtree agent game trees situations have a compatible generic diagram to the node.

Proof: Follows from game tress node satisfiability, definitions 4.3, 4.4, and 4.6.

Proposition 4: Agent game tree generic diagrams encode compatibility on satisfying the VMK game tree criteria for the node player.

Proof: Follows from VMK definitions 4.5 and 4.6, and proposition 4.3.

Proposition 5: The information partition for each player I at a node is $\bigvee$ D<fI > where D<fI > is the diagram definable to the player I's agent functions at that node, where $\bigvee$ is the infinite conjunction.

Proof: Since VMK has countable agent assignments at each game tree node there is a diagram 13 definable on the partition where countable conjunction can determine the next move. □

Theorem 4: A game tree is solved at a node iff there is an elementary embedding on the generic diagrams, sequentially upward to the node, corresponding to VMK played at the node, that solves the root node.

Proof: Elementary embedding has upward compatibility on tree nodes as a premise. Proposition 3.4 grants us that a VMK node check is encoded on diagram compatibility. Applying proposition 3.1, we can base diagram check the ordinal assignments in isomorphic embedding to compute a countable conjunction, hence an ordinal. On proposition 3.5, we have that compatibility is definable on a countable conjunction over the agent partition diagrams. Therefore, at the root node, a game goal formula φ is solved, iff at each game tree node to the terminal node the D<fI> on the nodes, for every model Mj, and a game sequence <S>, Msj |= D<fI> iff Msj–1 |= D<fI> on the ordered sequence s ∈ <H x I>, j an ordinal < |S|.

First author characterizes agent morphisms, in for example (Nourani 2005), to compute agent state- space model computations. So we can, for example, state the following.

Corollary 4.1: The agent game trees nodes at VMK are solved iff at the tree game sequence there are strong holomorphic game tree model diagram embeddings towards a terminal nodes where a goal diagram is satisfied.

From Proposition 3 we can address on a newer publications how model epistemic can be treated. The relationship to the diagram is an information set, an agent's knowledge is characterized by the diagram. What we have not stated is whether we can use modal logic or not. For starters, the diagram is no longer finite unless you make further assumptions, because you have to model what agents know and believe about each other. The above proposition allows us to treat infinite conditions. Model diagrams were treated on the first author's publications, for example, (Nourani, 2000).

D. Game Tree Degrees, Model Diagrams, and Payoff's: From the VMK game definition we have the following criteria

(a) For each player $i \in N$ an *information partition* Ii defined on $\{s \in H: \text{player}(s) = i\}$. An element Ii of Ii is called an *information set* of Player i. We require that if s, s′ are members of the same information set Ii, then $A(s) = A(s')$.
(b) For each player $i \in N$ a *payoff function* $u_i: Z \rightarrow R$ that assigns a real number to each terminal history.

Recall that on condition 5 at VMK definition 3.4 the information partition for a player I is determined at each game node by the game diagram game tree degree (Section 3.1) at that node that is known to that player. Based on Definition 3.5 and consequences, e.g., on (a,b) above, at VMK definition 3.4 the information partition for a player I is determined at each game node by the game diagram game tree degree (Section 4.1) at the node that is known to that player. Applying Proposition 4 and 5 with considerations that generic model diagrams with VMK game function situation, Definition 4.5, we have second-order lift on payoff functions. That is the measures on the nodes are game trees determined based on how the agent model sequent functions are satisfying the model diagrams at game tree node on Theorem 4. The measures are not lost, whatever the measures are, but only embedded on the game tree lifting preorders. The measures can be created on random sequences for all we know, for example.

7.4.1 TOPOLOGICAL SPACES AND GAMES

Let A, B be two topological spaces. A mapping $f: A \rightarrow B$ is *continuous* if for every open set $Y \subseteq B$, its preimage $f-1(Y)$ is an open subset of A. Thus, to define the set of continuous payoff functions, we need to introduce a system of open sets on Z, the set of infinite histories, and R, the set of real numbers.

We shall employ the standard topology on R, denoted by R. The *Baire topology* is a standard topology for a space that consists of infinite sequences, such as the set of infinite game histories.

It is important in analysis, game theory. Let A be a set of actions. Let $[s] = \{h: s \subset h\}$ be the set of infinite action sequences that extend s. The basic open sets are the sets of the form [s] for each finite sequence (situation) s. An open set is a union of basic open sets, including again the empty set ∅. We denote the resulting topological space by B(A).

For each rational q and natural number n > 0, define an open interval O(q,n) centered at q by O(q, n) = {x: |x − q| < 1/n}. Let Eu(q, n) = u−1(O(q, n)). Since u is continuous, each set Eu(q, n) is an open set in the Baire space and hence a union of situations. So the function u induces a characteristic relation *interval u(s,* q, n) that holds just in case [s] ⊆ Eu(q, n). In other words, *interval u(s,* q, n) holds iff for all histories h extending s the utility u(h) is within distance 1/n of the rational q. In the example with the discontinuous payoff function above, *interval u(s,1,n)* does *not* hold for any situation s for n > 1.

From Schulte et al. (2005) on a given any situation s at which the disaster has not yet occurred, there is a history h ⊃ s with the disaster occurring, such that u(h) = 0 < |1 − 1/n|. Intuitively, with a continuous payoff function u an initial situation s determines the value u(h) for any history h extending s up to a certain "confidence interval" that bounds the possible values of histories extending s. As we move further into the history h, with longer initial segments s of h, the "confidence intervals" associated with s become centered around the value u(h).

7.4.2 GAMES AND TOPOLOGIES

Beginning from (Schulte-Delgrande, 2004), let us examine the following: A successor operation +: *action* → *action* that takes an action **a** to the "next" action **a+**. We write **a(n)** as a shorthand for **a0+⋯+** where the successor operation is applied n times to a distinguished constant **a0** (thus **a(0)** = **a0).** The constants **a(n)** may serve as names for actions. As in the case of finitely many actions, introduce unique names axioms of the form **a(i)**⊭ **a(j)** where i⊭ j. The following induction axiom is examined.

$$\text{Axiom: } \forall P.[P(a0) \wedge \forall a.P(a) \rightarrow P(a+)] \rightarrow \forall a.P(a).$$

Lemma 1 (OSD). Let M = ⟨actions, +, ⌈ ⌉⟩ be a model of axiom and the unique names axioms. Then, for every a ∈ *actions,* there is one and only one constant **a(n)** such that ⌈a(n)⌉ = a.

Consider ⌈ ⌉ that is a 1–1 and total assignment of actions in the game form A(F) to action constants.

Theorem (OSD), Let A be a set of actions, and let u: B(A)→R be a continuous function. Let *payoff* be the axiom ∀h, n.∃s h. interval u(s, u(h), n). Then,

1. u satisfies *payoff,* and
2. if u′: B(A) → R satisfies *payoff,* then u′ = u.

From Section 3.4 given two structures and of the same signature Σ, a Σ-homomorphism is a map that can preserve the functions and relations. A Σ-homomorphism h is called a Σ-embedding if it is one-to-one and for every *n-ary* relation symbol *R* of Σ and any elements ai,…,an, the following equivalence holds: R(a1…,an) iff R(h(a1).h(an)). Thus, an embedding is the same thing as a strong homomorphism which is one-to-one.

For example, applying Corollary 3.1 to the above we have

Theorem 5: Let A be a set of actions on M, and let u: B(A)→R be a continuous function. Then, u satisfies a *payoff* axiom iff u can be extended to a strong holomorphic embedding on the M model diagram towards a terminal node satisfying a goal on M (Nourani, Schulte).

Applying the VMK theorem 4 and noting that every elementary embedding is a strong homomorphism we carry on to the following.

Theorem 6: Let A be a set of actions on M, and let u: B(A)→R be a continuous function. The u satisfies a *payoff* axiom at M iff there is an elementary embedding on the generic diagrams for M, sequentially upward to the node, corresponding to VMK played at the node that solves the root node state is for an agent. □

Considering epistemic accessibility based on the correspondence between possible worlds and situations encoded on generic diagrams (Nourani, 1994). We can state the following proposition.

Proposition 6: Let A be a set of actions on M, and let u: B(A)→R be a continuous function. The u satisfies a *payoff* axiom iff there is a generic diagram on M that is epistemically accessible by A on M that satisfies the payoff axiom at least on one game tree node.

7.4.3 COMPLETENESS ON GAME LOGIC AND NASH CRITERIA

We can state preliminary theorems on VMK agent games. Basic agent logic soundness and completeness areas were examined by the first author on Nourani (2002). Making preliminary assumptions on VMK game situations let us examine soundness and completeness questions. Let us consider stratification as the process whereby generic diagrams are characterized with recursive computations on agent functions on game trees, for example compare to a notion on Geneserth (2010). We can further consider backward chaining on game trees based on standard AI and Schulte (2003) as the premises.

Theorem 7: There is a sound and complete agent logic on VMK game situations provided:

(i) VMK agent language is a countable fragment.
(ii) The agent information partitions are definable on a countable generic diagram,
(iii) Game tree node ordinal is definable with a countable conjunction on the generic diagrams (Nourani, 2014).

Proof outline: Follows from theorem 4, the propositions, and by considering carrying soundness on forward model diagram stratifications and the embedding propositions above. Completeness is ascertained on upward model diagram compatibility, and compactness. To have a feel for that let us consider stratification as the process whereby generic diagrams are characterized with recursive computations on agent functions on game trees; for example, compare to a notion on Genesereth (2010). We can further consider backward chaining on games tree based on standard AI and Schulte (2003).

From the above, we can state theorems on Nash criteria based on the generic diagram characterizations.

7.4.4 DESCRIPTIONS, GAMES, AND TRACTABILITY

From the second author's briefs we might ponder how we have computability on game descriptions based on model diagrams on game node trees, since having a sound and complete logic can ascertain when and how the criteria are satisfied. Given that every action in M has a unique action of constant naming. It follows from the axioms that every situation in M has a unique situation constant naming it (as before, we may write *name M(s)),* which implies that there is a 1–1 onto mapping between the situations in M and $A<\omega$. The criteria accomplished are important on model characterizations. Looking back to theorem 2 and the correspondence of generic model diagrams encoding for possible worlds (Nourani 1994, 1998, 2009), we can contemplate the following.

Proposition 7: A<ω has a unique encoding on generic model diagrams where game model compatibility can be characterized and the computability questions addressed on game models (Nourani, 2015).

From the first author's descriptive computing publications we have the following tractability consequences on game tree model diagram computations. We define descriptive computation to be computing with G-diagrams for the model and techniques for defining models with G-diagrams from the syntax of a logical language. G-diagrams are diagrams definable with a known function set. Thus, the computing model is definable by generic model-diagrams with a function set. The analogous terminology in set theory refers to sets or topological structure definable in a simple manner. Thus, by descriptive computation we can address artificial intelligence planning and theorem proving, for example.

The latter computational issues are pursued by the first author in Nourani (1984, 1992, 2010). The logical representation for reaching goals in general might be infinitary only. The first author in Nourani (1994a,b, 1996) shows that the artificial intelligence problem is to acquire a decidable descriptive computation for the problem domain. The infinite tree premise is a condition in (Schulte & Delgrande, 2004) on continuous payoff characterizations as well. Hence logic on infinitary games was applied in our publications. The first author (Nourani, 1996) proves two specific theorems for descriptive computing on diagrams. A compatibility theorem applies descriptive computing to characterize situation compatibility. Further, a computational epistemic reducibility theorem is proved by the descriptive computing techniques on infinitary languages by the first author. A deterministic epistemic is defined and it is proved not reducible to known epistemic. We further define apply infinitary logic and cardinality with transitive closure properties on sets and languages to define descriptive computable and admissible sets. We have defined a set to be descriptive computable, if it is definable by a G-diagram with computable functions and proved. What is accomplished above is that (A) for descriptive computable sets the interesting transitive closures on sentences are is definable from a generic-diagram by recursion. (B) For A an admissible computable set, A is descriptive computable. Cardinality restrictions on concepts are important areas explored by AI. The concept description logics systems allow users to express local cardinality on particular role filers. Global restrictions on the instances of a concept are difficult and not possible. Cardinality restrictions on concepts can be applied as an application domain description logic (Baader et al., 2003). The concept definitions with generic diagrams for localized KR and its relations to descriptive computable sets can be applied to concept cardinality restriction. By applying localized functions to define generic diagrams models for languages can be generated with cardinality restrictions.

7.5 PROOF TREES ON GAMES

7.5.1 GAME TREE COMPUTATIONAL HEURISTICS AND STRATIFICATION

Let us now view the deductive methods, for example, the proof-theoretic example: SLDNF-resolution, a well-known deductive heuristic. A SLDNF-proof can be considered as the unfolding of an AND/OR-tree, which is rooted in the formula to be proven, whose branches are determined by formulas of the theory, and whose leaves are determined by atomic formulas that are true in a world. Partial deduction from our viewpoint (Nourani-Hoppe, 1995) usually computes from a formula and a theory an existential quantified diagram. In these papers and Nourani (1995, 2005), we also instantiate proof tree leaves with free Skolemized trees, where free trees are substituted for the leaves. By a free Skolemized tree we intend a term built with constant symbols and Skolem functions terms. Dropping the assumption that proof-tree leaves get instantiated with atomic formulas only yields an abstract and general notion of proof trees. The mathematical formalization that allows us to apply the method of free proof trees is based on the first author's 1995–2005. In the present approach, as we shall further define, leaves could be free Skolemized trees. By a free Skolemized tree we intend a term made of constant symbols and Skolem function terms. Like models and diagrams, which where generalized above in different ways, we can generalize the notion of a proof. First author had developed free proof tree techniques since projects at TU Berlin, 1994. Free proof trees allow us to carry on Skolemized tress on game tree computing models, for example, that can have unassigned variables. The techniques allow us to carry on predictive model diagrams. Reverse Solemnization (Nourani, 1986) that can be carried on with generic model diagrams corresponding to what since Genesereth (2011) is applying on game tree "stratified" recursion to check game tree computations. The free trees defined by the notion of provability implied by the definition, could consist of some extra Skolem functions {g1,...,gm},that appear at free trees. The *f* terms and *g* terms, tree congruencies, and predictive diagrams then characterize partial deduction with free trees. To compare recursive stratification on game trees on what Genesereth calls recursive stratification we carry on models that are recursive on generic diagram functions where goal satisfaction is realized on plans with free proof trees (Nourani, 1994–2007). Thus, essentially the basic heurstics here is satisfying nodes on agent AND/OR game trees. The general heuristics to accomplish that are a game tree deductive technique based on computing game tree unfoldings projected onto predictive model diagrams. The soundness and completeness of these techniques, e.g., heuristics as a computing logic is published since 1994 at several events (Berlin logic colloquium, Potsdam Universitat Mathematik, 1994; AISB, 1995; Systems and Cybernectics, 2005). The heuristic nomenclature indicates that a heuristic function is called an *admissible- heuristic* if it never overestimates the cost of reaching the goal, i.e. the cost it estimates to reach the goal is not higher than the lowest possible cost from the current point in the path. An admissible heuristic is also known as an *optimistic heuristic* (Russell and Norvig, 2002). We shall use the term optimistic heuristic from now on to save ambiguity with admissible sets from mathematics (Shoenfield, 1967, or the first author's publications on descriptive computing, for the time being. What is the cost estimate on satisfying a goal on an unfolding projection to model diagrams, for example with SLNDF, to satisfy a goal? Our heuristics are based on satisfying nondeterministic Skolemized trees (Nourani 1994). The heuristics aim to decrease the unknown assignments on the trees. Since at least one path on the tree must have all assignments defined to T, or F, and at most one such assignment closes the search, the "cost estimate," is no more than the lowest. Let us call that nomenclature as one compound word: "admissible heuristics," not to confuse that with the notion of admissible sets at mathematical logic in the first author's publications. We do not know if we can have a relationship to that mathematical area yet. To become more specific how game tree node degrees can be ranked, we state one example linear measure proposition.

Proposition 8: Supposing the degree rank is defined based on a linear counting measure-the intersection at a goal state, where there are true literals, counting the goal literals that become true, assuming that each action can make at most one goal literal true at a time, then the ranking is admissible heuristic ranking function (Nourani, 2015).

Not being restricted to the above proposition on specific ranks, only that there is viable node degree rank, we have more general areas to address as follows, independent of the specific ranking notions.

Lemma 2: AND/OR agent game tree satisfaction with unfolding projections onto predictive model diagrams is an optimistic heuristic.

We can prove that direct deploying more on models, or carry on with two corollaries. The process to reach an admissible heuristic based on SLDNF is done with defining a measure on what literals are selected first on assignments. The process entails that the minimal route to resolving literals are chosen so as to accomplish an admissible heuristic with SLDNF. The measures are defined in a precise mathematical manner (Ntienjem, 1997).

Corollary 6.1: SLNDF deductive heuristics based on admissible selection functions is an optimistic heuristic.

Proof: Follows from a report on SLNDF computing logic at the Universitat Augsburg (Ntienjem, 1997).

Corollary 6.2: Predictive diagram unfolding encodes SLDNF techniques on partial deductions.

Proof: Nourani-Hoppe (1995) TU Berlin, Potsdam Universitat Logic colloquium, AISB (1995), and Nourani (2005) Systems and Cybernetics.

7.5.2 GAME NODE DEGREE HEURISTICS EXAMPLE

A Nash equilibrium (NE) is a joint strategy such that no agent may unilaterally change its strategy without lowering its expected payoff in the one shot play of the game. Nash (1951) showed that every n-player matrix game has at least one. Each node has a heuristic value which says how close a given state is to the goal state. With these values you can determine which node is the best to move to next. In a turn-based games like chess you can always calculate the exact number of possible states. The current state will not change until you do something. In a game that is not turn-based your opponent can change the current state at any given time. Since the moment is random the number of possible states can be considered infinite. In a non-turn based game tree you also can't have heuristic values based on closeness to goal. In a non-turn based game tree there is no point in looking far ahead because the current state can change radically at any moment. The heuristic algorithm must not only incorporate the closeness to goal but also estimate how the situation could change due to actions of the opponent. Let us consider examples towards specific game tree computations. Suppose a goal is described by a DNF formula G1 or G2 … or Gn.

An example linear degree assignment is that the node degree is the max over the disjuncts of the intersections. For example, if goal formula Gi shares k literals with the literals that are known to be true at node u, then the degree of u is at least k. Let us now consider an example that was run at (Schulte & Delgrande, 2004). Consider servers on the internet. Each server is connected to several sources of information and several users, as well as other servers. There is a cost to receiving and transmitting messages for the servers, which they recover by charging their users. We have two servers Srv1 and Srv2, and two users – journalists – U1 and U2. Both servers are connected to each other and to user U1; server Srv2 also serves user U2. There are two types of news items that interest the users: politics and showbiz. The various costs and charges for transmissions add up to payoffs for the servers, depending on what message gets sent where. For example, it costs Srv1 4 cents to send a message to Srv2, and it costs Srv2 2 cents to send a message to U2. If U2 receives a showbiz message from Srv2 via Srv1, he pays Srv1 and Srv2 each 6 cents. So in that case the overall payoff to Srv1 is −4 + 6 = 2, and the overall payoff to Srv2 is −2 + 6 = 4. Schulte & Delgrande (2005) describes the specifics in detail; a summary is given in Figure 7.4.

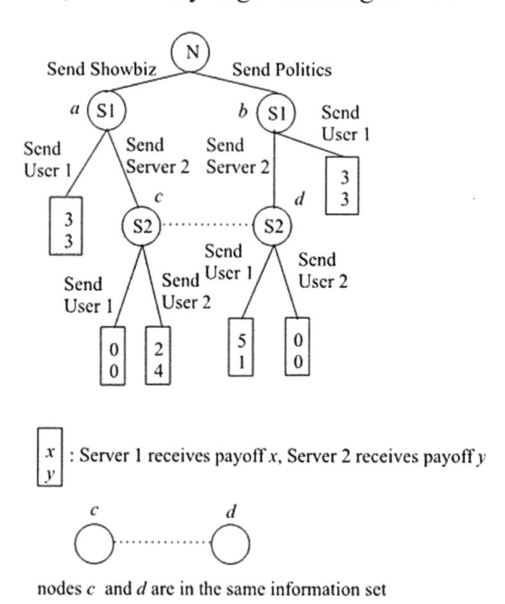

FIGURE 7.4 A game-theoretic model of the node diagram. (a) Action(N) = Showbiz; (b) Action(N) = Politics; (c) Action(S1) = Send Server 2, Action(N) = Showbiz OR Action(N) = Send Politics; (d) Action(S1) = Send Server 2, Action(N) = Showbiz OR Action(N) = Send Politics.

Table 7.1 The Model Diagrams for Each Choice Node in the Game Tree

Node	Diagram
a	Action (N) = Showbiz
b	Action (N) – Politics
c	Action (S1) = Send Server 2,
	Action (N) = Showbiz OR Action (N) = Send Politics
d	Action (S1) – Send Server 2,
	Action (N) = Showbiz OR Action (N) = Send Politics

For example for *f* and *g* defining agents for two Boolean valued relations on a decision tree the diagram specifies the Boolean values based on parameter assignments to X and Y.

7.5.3 MATRIX AGENT GAMES AND NASH EQUILIBRIUMS

Examining Multiagent Reinforcement Learning for iterated play of games, more specifically matrix games, let us introduce some well-known concepts from game theory. In general, let S denote the set of states in the game and let Ai denote the set of actions that agent/player i may select in each state s ϵ S. Let a = (a1, a2,..., an), where ai ϵ Ai be a joint action for n agents, and let A = A1 x...x An be the set of possible joint actions. Zero-sum games are games where the rewards of the agents for each joint action sum to zero.

General sum games allow for any sum of values for the reward of a joint action. A strategy for agent i is a probability distribution π(.) over its actions set Ai. Let π(S) denote a strategy over all states s ϵ S and let π(s) (or πi) denote a strategy in a single state s. One way to look at equilibrium on diagrams is that we consider.

An n-player game on a matrix, where each row corresponds to a player diagram.

We focus on the more restricted matrix game, denoted by a set of matrices R = R1,...Rn. Let R(π) = (R1(π), ..., Rn(π)) be a vector of expected payoffs when the joint strategy π is played. Also, let Ri (π i, π–i) be the expected payoff to agent i when it plays strategy πi and the other agents play π–i.

A stage game is a single iteration of a matrix game, and a repeated game is the indefinite repetition of the stage game between the same agents. While matrix games do not have state, agents can encode the previous w joint actions taken by the agents as state information. Each individual matrix game has certain classic game theoretic values. The minimax value for player *i* is the least reward that can be achieved if the game is known and the game is only played once. There is a natural strategy based on a single player's view to *n*-player game that does not have to have a paring characterization. Nash equilibrium naturally occurs based on how well each player heuristic on SLNDF for example unfolds. Based on the above we can have a proposition.

Proposition 9: A matrix game reaches Nash equilibrium when all the player corresponding model diagram rows are defined and at least one player accomplishes the goal set(s) (Nourani & Schulte, 2015).

7.5.4 MORE ON INTERACTIVE DECISIONS

Games theory is theory of rational behavior for interactive decision problems. In a game, several agents strive to maximize their (expected) utility index by choosing particular courses of action, and each agent's final utility payoffs depend on the profile of courses of action chosen by all agents. The interactive situation, specified by the set of participants, the possible courses of action of each agent, and the set of all possible utility payoffs, is called a game. The agents 'playing' a game are called the players. In degenerate games, the players' payoffs only depend on their own actions. For example, in competitive markets, it is enough that each player optimizes regardless of the behavior of other traders. As soon as a small number of agents are involved in an economic transaction, however, the payoffs to each of them depend on the other agents' actions with the Shculte-Delgrande (2005) notation, for example. We can cast the interactive Nash based on the above sections. Let us define the payoff to each player that results when two strategies are paired in situation s, which is just their utility from the resulting action sequence:

$$\mathrm{Ui}(\pi 1, \pi 2, \mathrm{s}) = \mathrm{ui}$$
$$\mathrm{Play}\ (\pi 1, \pi 2, \mathrm{s})$$

For the notion of Nash equilibrium we need the concept of a best reply,

$$\text{best Reply } 1\ (\pi 1, \pi 2, \mathrm{s}) \equiv \forall \pi 1.\mathrm{valid}(\pi 1)) \rightarrow \mathrm{U1}(\pi 1, \pi 2, \mathrm{s})\ \mathrm{U1}(\pi 1, \pi 2, \mathrm{s}),$$
$$\text{best Reply } 2\ (\pi 1, \pi 2, \mathrm{s}) \equiv \forall \pi 2.\mathrm{valid}(\pi 2) \rightarrow \mathrm{U2}(\pi 1, \pi 2, \mathrm{s})\ \mathrm{U2}(\pi 1, \pi 2, \mathrm{s}).$$

A pair of strategies ($\pi 1$, $\pi 2$) forms a Nash equilibrium in a game overall just in case each strategy is a best response to the other. The payoff from two strategies in the overall game is just the result of starting in the initial situation

S0. Thus, we have, Nash (π1, π2) ≡ valid(π1) ∧ valid(π2) ∧ bestReply1(π1, π2, S0) ∧ bestReply2(π1, π2, S0).

7.5.5 PREDICTIVE PLAN PROOF TREES

The chapter accomplishes modeling a multiplayer cooperative game to reach a goal or modeling group action game where group A is playing group B where B's moves are not apriori known. Neither is both sides information set. Hence the information set that a group player carries on to base moves on, is changed on a diagram. The models and encoding apply to zero-sum AND/OR arbitrary games. We have viewed the game as zero-sum on a game from a group action stage that plays to reach a goal where certain propositions are true. New VMK encodings allow us to reach new mathematical goals on game model computability towards specific Nash equilibrium models. A novel basis to decision-theoretic planning is presented classical and non-classical planning techniques, see for example (Hedeler et al., 1990; Wilkins, 1984) from artificial intelligence with games and decision trees providing a agent expressive planning model following the first authors competitive game model planning publication can be developed based on the new techniques here. We use a broad definition of decision-theoretic planning that includes planning techniques that deal with all types of uncertainty and plan evaluation. Planning with predictive model diagrams represented with keyed KR to knowledge bases is presented. Techniques for representing uncertainty, plan generation, plan evaluation, plan improvement, and are accommodate with agents, predictive diagrams, and competitive model learning. Modeling with effector and sensor uncertainty, incomplete knowledge of the current state, and how the world operates can be treated with agents and competitive models. Bounds on game trees can be stated based on the first author's preceding publications agent games and the second author's generalizations on VMK to agents. The heuristics area can be further developed based on a comprehensive treatment on game unfolding and computability criteria on model diagrams. Computability and reachability on Nash models are further areas to explore.

7.5.6 PROOF TREES ON GAMS

The first author had developed free proof tree techniques since projects at TU Berlin, 1994. Free proof trees allow us to carry on Skolemized tress on game tree computing models; for example, that can have unassigned variables. The techniques allow us to carry on predictive model diagrams. Reverse Solemnization (Nourani, 1986) that can be carried on with generic model diagrams might correspond to what since Generserth (2010) is applying on game tree "stratified" recursion to check game tree computations. Existentially quantified goals on diagrams carry a main deficit. The Skolemized formulas are not characterized. A *predictive diagram* for a theory T is a diagram D[M], where M is a model for T, and for any formula q in M, either the function f: q → {0,1} is defined, or there exists a formula p in D[M], such that T U {p} proves q; or that T proves q by minimal prediction. A *generalized predictive diagram,* is an predictive diagram with D[M] defined from a minimal set of functions. The predictive diagram could be minimally represented by a set of functions {f1,...,fn} that inductively define the model. The free trees [11] defined by the notion of provability implied by the definition, could consist of some extra Skolem functions {g1,...,gl}that appear at free trees. The *f* terms and *g* terms, tree congruences, and predictive diagrams then characterize partial deduction with free trees. To compare recursive stratification on game trees on what Geneserth calls recursive stratification we carry on models that are recursive on generic diagram functions where goal satisfaction is realized on plans with free proof trees (Nourani, 1994–2007).

Theorem 8: For the free proof trees defined for a goal formula from a generic diagram there is a canonical model satisfying the goal formulas. It is the canonical initial model created with the generic diagram (Nourani, 2005).

Proof: In planning with generic diagrams plan trees involving free Skolemized trees is carried along with the proof tree for a plan goal. The idea is that if the free proof tree is constructed then the plan has a model in which the goals are satisfied. There is analogy to SLD proofs. We can view on the one hand, SLD resolution type proofs on ground terms, where we go from p(0) to p(f(c)); or form p(f(c)) to p(f(g(c)). Whereas, while doing proofs with free Skolemized trees we are facing proofs of the form p(g(....)) proves p(f(g(....)) and generalizations to p(f(x)) proves For all x , p(f(x)). Since the proof trees are either proving plan goals for formulas defined on the G-diagram, or are computing with Skolem functions defining the G-diagram, by generic diagram definition, and theorems on Nourani (1994) the model defined by the generic diagram applies and it is canonical, e.g., Initial, for the proofs.

7.6 CONCLUSION AND AREAS TO EXPLORE

Modeling a multiplayer cooperative game to reach a goal or modeling group action game where group A is playing group B where B's moves are not apriori known. Neither is both sides information set. Hence the information set is changed on a diagram, that A group player carries on to base moves on. The models and encoding apply to zero-sum AND/OR arbitrary games. We have viewed the game as zero- sum on a game from a group action stage that plays to reach a goal where certain propositions are true. A novel basis to decision-theoretic planning is presented classical and non-classical planning techniques, see for example, Hedeler et al. (1990) and Wilkins (1984) from artificial intelligence with games and decision trees providing a agent expressive planning model. We use a broad definition of decision-theoretic planning that includes planning techniques that deal with all types of uncertainty and plan evaluation. Planning with predictive model diagrams represented with keyed KR to knowledge bases is presented. Techniques for representing uncertainty, plan generation, plan evaluation, plan improvement, and are accommodate with agents, predictive diagrams, and competitive model learning. Modeling with effector and sensor uncertainty, incomplete knowledge of the current state, and how the world operates is treated with agents and competitive models. Bounds on game trees can be stated based on the first author's preceding publications agent games and the second author's generalizations on VMK to agents.

KEYWORDS

- **analytics support systems**
- **competitive model computing**
- **game description logic**
- **game sequent description**
- **game tree models**
- **Nash games**
- **VMK games**

REFERENCES

Badder, F., Buchheit, M., & Hollunder, B., (1996). *Cardinality Restrictions on Concepts*, AI.

Brazier, F. M. T., Dunin-Keplicz, B., Jennings, N. R., & Treur, J., (1997). DESIRE: modeling mulch- agent systems in a compositional formal framework. In: Huhns, M., & Singh, M., (eds.), *International Journal of Cooperative Information Systems* (Vol. 1). Special issue on formal methods in cooperative information systems, electronic edition.

Breiman, L., (1996). Some properties of splitting criteria. *Machine Learning, 24*, 41–47.

Breiman, L., Friedman, J. H., Olshen, R. A., & Stone, C. J., (1984). *Classification and Regression Trees*. Chapman & Hall (Wadsworth, Inc.): New York.

Chung, C. C., & Kiesler, H. J., (1973). *Model Theory*, Elsevier, ISBN 978-0-7204-0692-4.

Franz Baader et al., (2003). *Theory, Implementation, and Applications*. The Description Logic Handbook. Aachen University of Technology, ISBN: 9780521781763.

Gale, D., & Stewart, F. M., (1953). "Infinite games with perfect information," in contributions to the theory of games. *Annals of Mathematical Studies, 28*, Princeton.

Genesereth, M. R., & Nilsson, N. J., (1987). *Logical Foundations of Artificial Intelligence*, Morgan Kaufmann.

Genesereth, M., (2010). *Stanford University Computer Science Lecture Notes on Games.*

Goguen, J. A., Thatcher, J. W., Wagner, E. G., & Wright, J. B., (1973). *"A Junction Between Computer Science and Category Theory,"* (Parts I and II). IBM T.J. Watson Research Center, Yorktown Heights, N.Y. Research Report, RC4526.

Holland, J. H., & Miller, J. H., (1991). Artificial adaptive agents in economic theory, American Economic Association. *Journal American Economic Review*. Issue no. 81.

Kendall, D. G., (1974). Foundations of a theory of random sets. In: Harding, E. F., & Kendall, D.G., *Stochastic Geometry* (pp. 322–376). John Wiley, New York.

Kleene, S. A., (1951). *"Introduction to Metamathematics."* North Holland.

Koller, D., & Pfeffer, A., (1997). "Representations and solutions for game-theoretic problems." *Artificial Intelligence, 94*(1), 167–215.

LOFT, (2008). Logic, and the foundations of game and decision theory – 8th international conference, Amsterdam, The Netherlands. In: Bonanno, G., Löwe, B., & Van der Hoek, W., (eds.), *Series: Lecture Notes in Computer Science* (1st edn., Vol. 6006). Subseries: Lecture Notes in Artificial Intelligence, 2010, XI, 207 p ISBN: 978–3–642–15163–7.

Nash, J. F., (1951). Non-cooperative games. *Annals of Mathematics, 54*(1951), 286–295.

Nash, J., (1950). Equilibrium points in n-person games. *Proc. National Academy of Sciences of the USA, 36*, 48, 49.

Nilsson, N. J., (1969)."Searching, problem solving, and game-playing trees for minimal cost solutions." In: Morell, A. J., (ed.), *IFIP 1968* (Vol. 2). Amsterdam, North Holland, 1556–1562.

Nilsson, N. J., (1971). *Problem Solving Methods in Artificial Intelligence."* NewYork, McGraw-Hill.

Nourani, C. F. *"Higher Stratified Consistency and Completeness Proofs,"* Summer logic colloquium, Helsinki, http://www.logic.univie.ac.at/cgi.-bin/abstract/show.pl?new=e049a2efe0c1a4b7a3ddaa11a75d8152, *Mathematicians*, Vol. II (Cambridge, 1913).

Nourani, C. F., (1984). "Equational intensity, initial models, and AI reasoning," Technical Report, (1983): A conceptual overview. In: *Proceedings Sixth European Conference in Artificial Intelligence*, Pisa, Italy, North Holland.

Nourani, C. F., (1991). "Planning and Plausible Reasoning in AI." *Proceedings Scandinavian Conference in AI, May,* Denmark, 150–157, IOS Press.
Nourani, C. F., (1994). "A theory for programming with intelligent syntax trees and intelligent decisive agents-preliminary report" 11th European Conference A.I., *ECAI Workshop on DAI Applications to Decision Theory*, Amsterdam.
Nourani, C. F., (1995). *"Free Proof Trees and Model-Theoretic Planning."* Automated Reasoning AISB, England.
Nourani, C. F., (1996a). Slalom tree computing – a tree computing theory for artificial intelligence. *AI Communications, European AI Journal* (Vol. 9, No. 4). IOS Press. 10.3233/AIC–1996–9402
Nourani, C. F., (1997). *"Descriptive Computing-The Preliminary Definition."* Summer logic colloquium, San Sebstian Spain. See AMS, Memphis.
Nourani, C. F., (1997). *"Multiagent Chess Games."* AAAI Chess Track, Providence, RI.
Nourani, C. F., (1997). *"Syntax Trees, Intensional Models, and Modal Diagrams, For Natural Language Models*. Uppsala logic colloquium, Uppsala University, Sweden.
Nourani, C. F., (1997). Intelligent Tree Computing, Decision Trees and Soft OOP Tree Computing, September 2, 1997, *Frontiers in Soft Computing and Decision Systems*, Papers from the 1997 Fall Symposium, Boston, Technical Report FS–97–04 AAAI, ISBN: 1–57735–079–0, www.aaai.org/Press/Reports/Symposia/Fall/fs-97-04.html.
Nourani, C. F., (1998). "Intelligent languages, a preliminary syntactic theory." In: Kelemenová, A., (ed.), *Proceedings of the Mathematical Foundations CS'98 Satellite Workshop on Grammar Systems* (pp. 281–287). Silesian University, Faculty of Philosophy and Sciences, Institute of Computer Science, Opava.
Nourani, C. F., (1999a). "Agent computing, KB for intelligent forecasting, and model discovery for knowledge management." *AAAI-Workshop on Agent Based Systems in the Business Context,* Orlando, Florida.
Nourani, C. F., (1999a). *"Infinitary Multiplayer Games,"* Utrecht. The Bulletin of Symbolic Logic, 2000. *European Summer Meeting of the Association for Symbolic Logic* (Vol. 6, No. 1).
Nourani, C. F., (2000). *Descriptive Definability: A Model Computing Perspective on the Tableaux.* CSIT 2000 Ufa, Russia. *de/~cp/p/zombie/.*
Nourani, C. F., (2001). *Management Process Models and Game Trees Applications to Economic Games*. Invited Paper SSGRR, L'Auquila, Rome, Italy.
Nourani, C. F., (2002). Game trees, competitive models, and ERP, new business models and enabling technologies, management school, St Petersburg, Russia, Keynote Address. June 2002, Nikolia Krivulin. Saint Petersburg State University, Russian Federation. *Fraunhofer Institute for Open Communication Systems, Germany, 11–12.* www.math.spbu.ru/user/krivulin/Work2002/Workshop.htm. Proceedings Editor Nikolai Krivulin.
Nourani, C. F., (2003). *Predictive Model Discovery, and Schema Completion.* World multiconference on systemics, cybernetics, and informatics (SCI 2002) Orlando, USA, July 14–18, 2003, http://www.iiis.org/sci2002.
Nourani, C. F., (2009). A descriptive computing, information forum, Leipzig, Germany. SIWN2009 Program, 2009. *The Foresight Academy of Technology Press International Transactions on Systems Science and Applications*, (Vol. 5, No. 1, pp. 60–69).
Nourani, C. F., (2012). *Competitive Models, Game Tree Degrees, and Projective Geometry on Random Sets – A Preliminary*. ASL annual meeting, Wisconsin.

Nourani, C. F., & Grace, S. L., & Loo, K. R., (2000). *Model Discovery From Active DB With Predictive Logic, Data Mining.* Applications to Business and Finance Cambridge, UK.
Nourani, C. F., & Hoppe, T., (1994). *GF-Diagrams for Models and Free Proof Trees, Technical Universitat Berlin, Informatiks, Berlin Logic Colloquium.* Universitat Potsdamm, Potsdamm, Germany. Humboldt Universtiat Mathematik sponsor.
Nourani, C. F., & Moudi, R. M. (2005). *Open Loop Control and Business Planning: A Preliminary.* Brief, International Conference Autonomous Systems, Tahiti, French Polynesia, (2005). www.iaria.org/conferences/ICAS/ICAS2005/General Information/ GeneralInformation.html, p. 8, ISBN: 0-7695-2450-8.
Ntienjem, E., (1997). "Completeness and Termination of SLNDF- Resolution and Determination of a Selection function using Mode." *Universitat Ausburg, Report* 1997–06.
Parikh, R., (1983/1998). Propositional game logic. In: Poole, D., (ed.), *IEEE Symposium on Foundations of Computer Science* (pp. 195–200). Decision theory, the situation calculus, and conditional plans, Linköping electronic articles in computer and information science, *3*(8).
Peter, F., & Patel-Schneider, (1990). A decidable first-order logic for knowledge representation. *Journal of Automated Reasoning, 6*, pp. 361–388. A preliminary version published as AI Technical Report Number 45, Schlumberger Palo Alto Research.
Russell, S. J., & Norvig, P., (2002). *Artificial Intelligence: A Modern Approach.* Prentice Hall. ISBN 0-13-790395-2.
Schulte, O., (2003). Iterated backward inference: An algorithm for proper rationalizability. *Proceedings of TARK IX (Theoretical Aspects of Reasoning About Knowledge)*, Bloomington, Indiana, pp. 15–28. ACM, New York. Expanded version with full proofs.
Schulte, O., & Delgrande, J., (2004). Representing von Neumann-Morgenstern Games in the Situation Calculus. *Annals of Mathematics and Artificial Intelligence, 42*(1–3), 73–101. (Special Issue on Multiagent Systems and Computational Logic). A shorter version of this paper appeared in the 2002 AAAI Workshop on Decision and Game Theory.
Shoenfield, J. R. (2001). Mathematical Logic (Paperback – February 9, 2001). ASL Publications. Hard cover 1967 Addison-Wesley Educational Publishers Inc. (December 1967) ISBN-10: 0201070286 ISBN-13: 978-0201070286.
The Knowledge Engineering Review, (2012). Special Issue 02 (Agent-Based Computational Economics). *Cambridge Online Journals*, 27(2).
Von Neumann & Morgenstern, O., (1994), *Theory of Games and Economic Behavior.* Princeton University Press, Princeton, NJ.
Vulkan, N., (2012). Strategic design of mobile agents, *AI Magazine, 23*(3), AAAI.
Wilkins, D., (1984). Domain-Independent Planning: Representation and Plan Generation. *AI, 22*(3), 269–301.

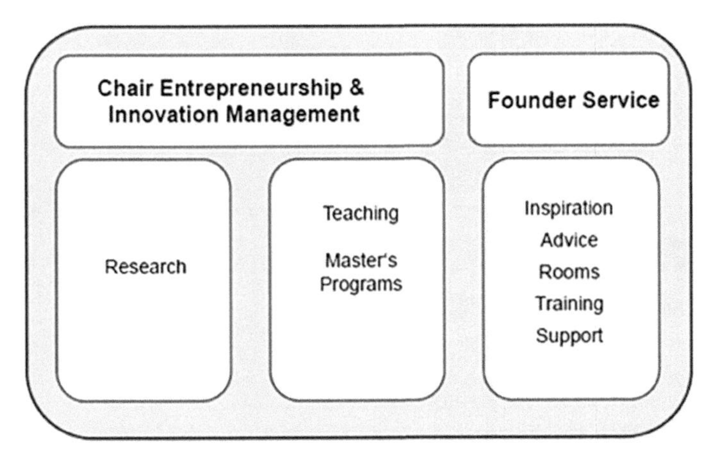

FIGURE 1.1 Organizational structure of center for entrepreneurship.

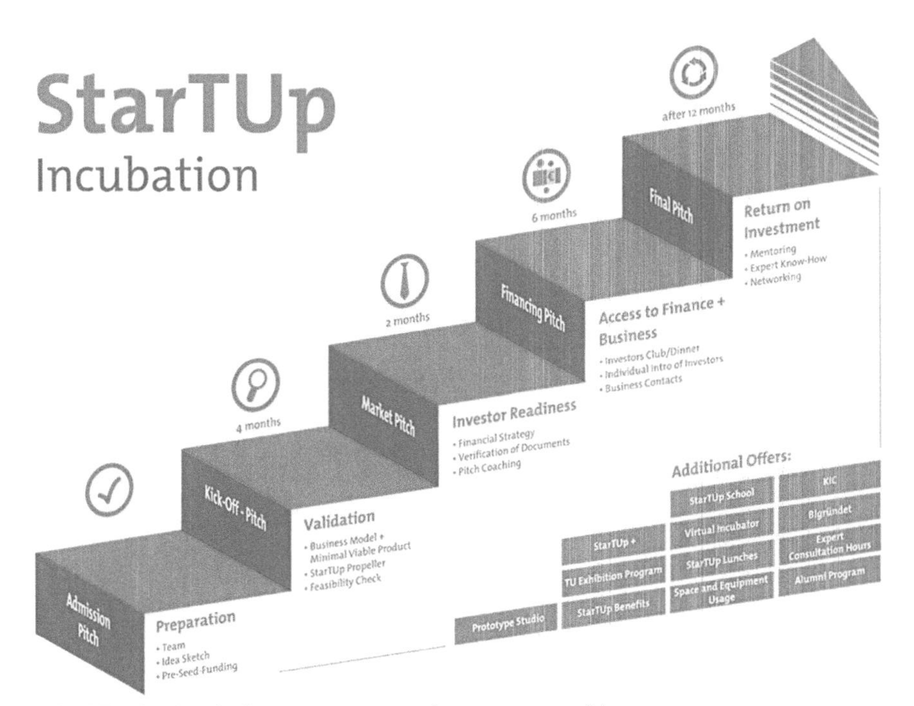

FIGURE 1.2 Incubation program center for entrepreneurship.

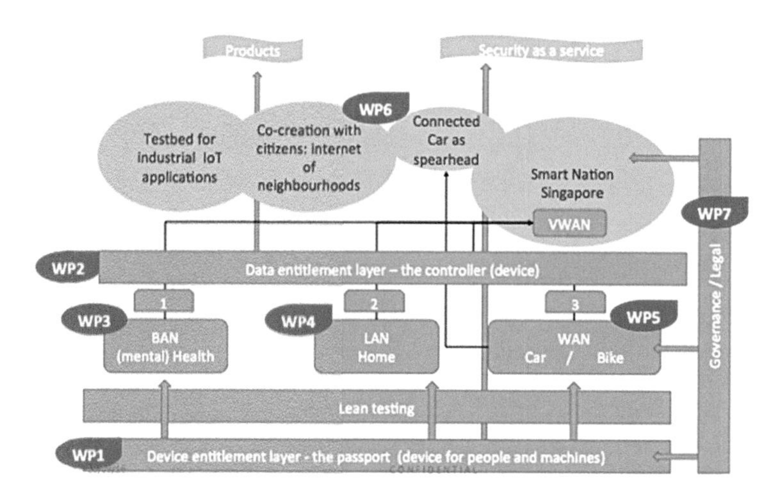

FIGURE 3.1 Technology.

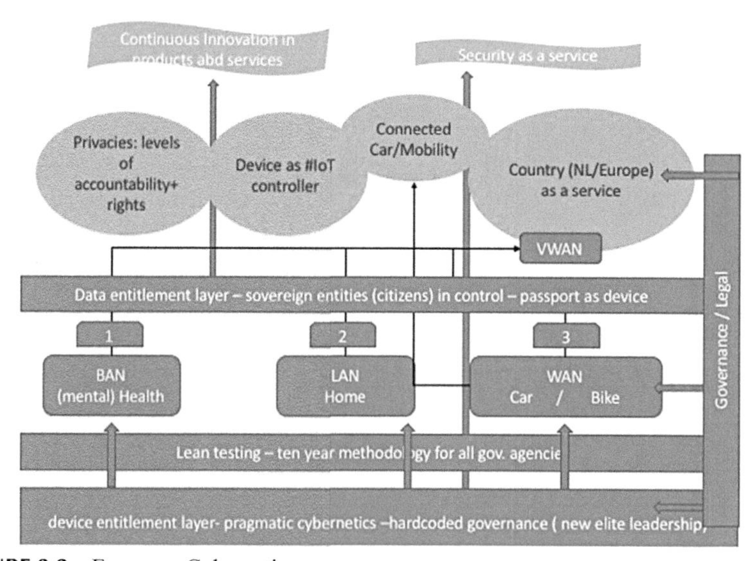

FIGURE 3.2 European Cybernetics.

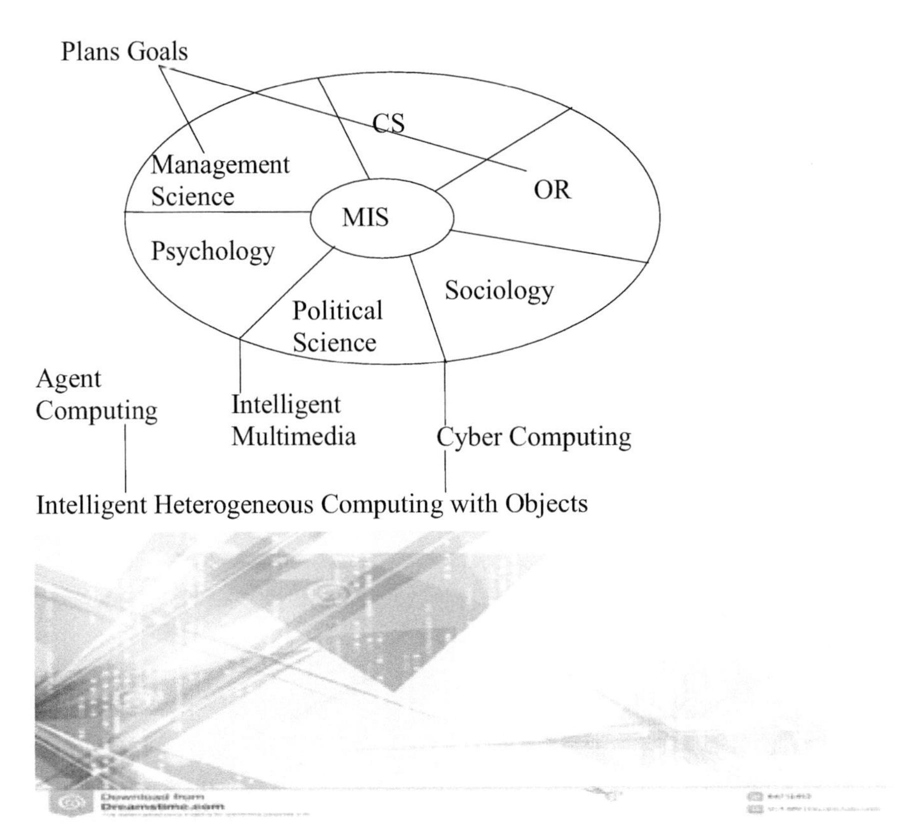

FIGURE 4.1 New MIS areas.

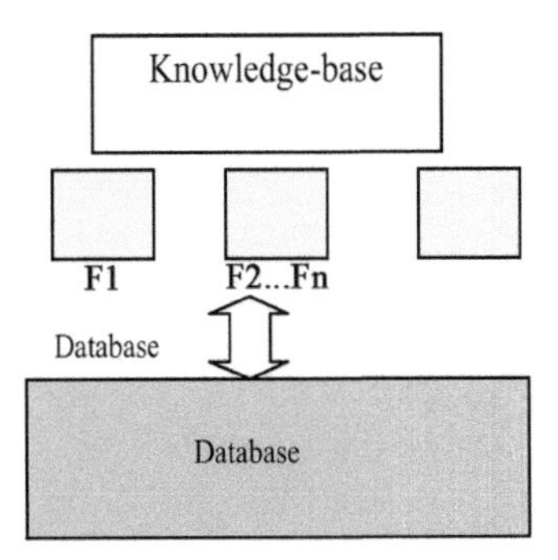

FIGURE 4.3 Keyed data functions, inference, and model discovery.

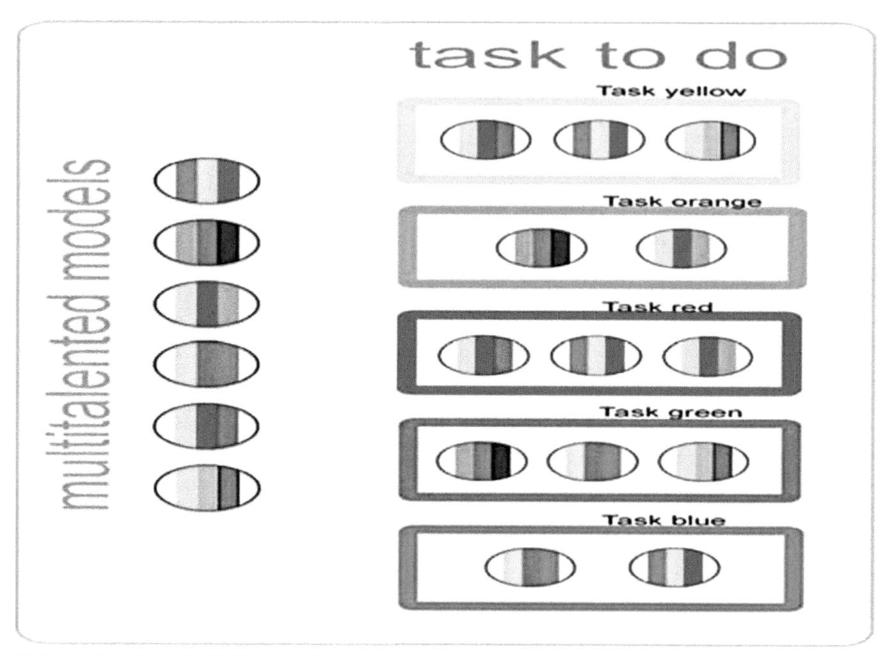

FIGURES 4.4, 7.1, 10.2 & 11.2 Competitive/multitalented models/games.

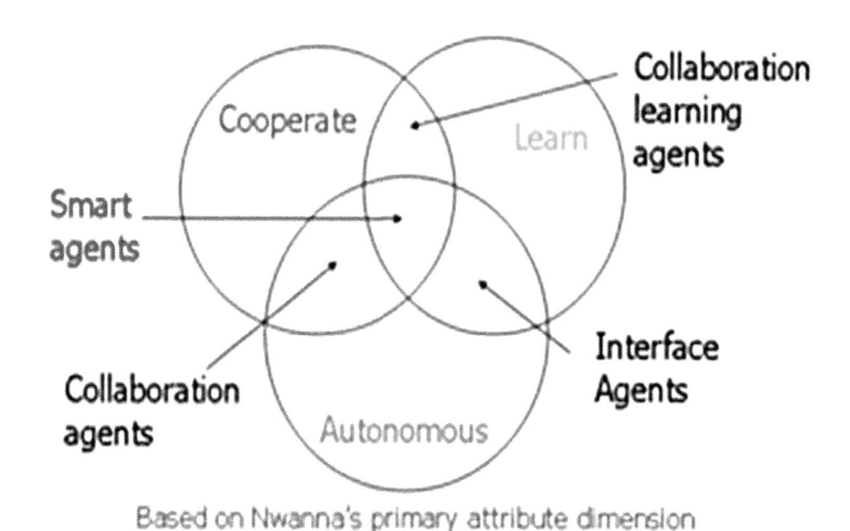

FIGURE 4.5 Cognitive cooperative learning.

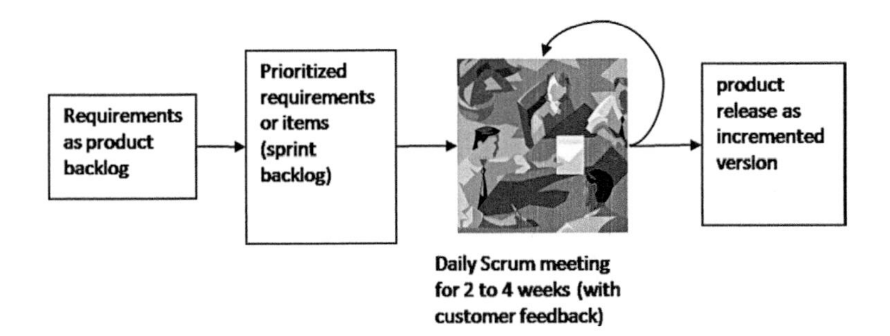

FIGURE 9.3 Scrum sprint cycle.

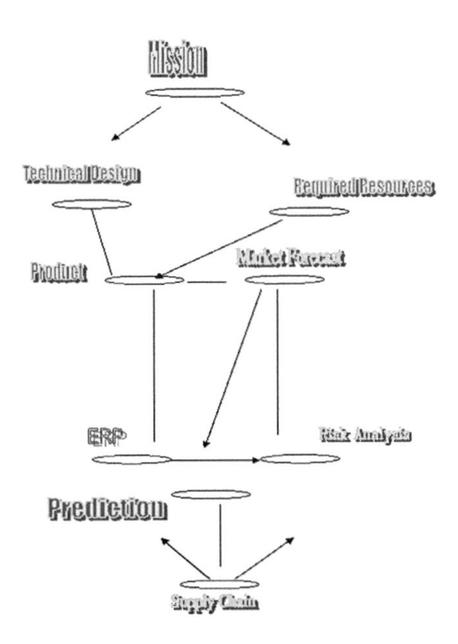

FIGURE 10.1 A business planning example.

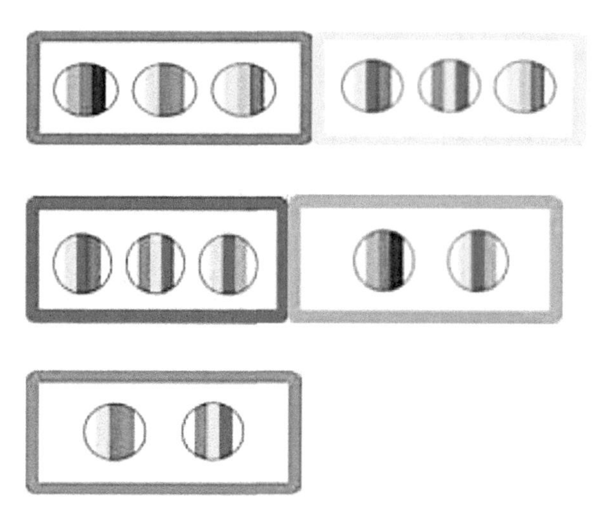

FIGURE 10.4 Task code optimality.

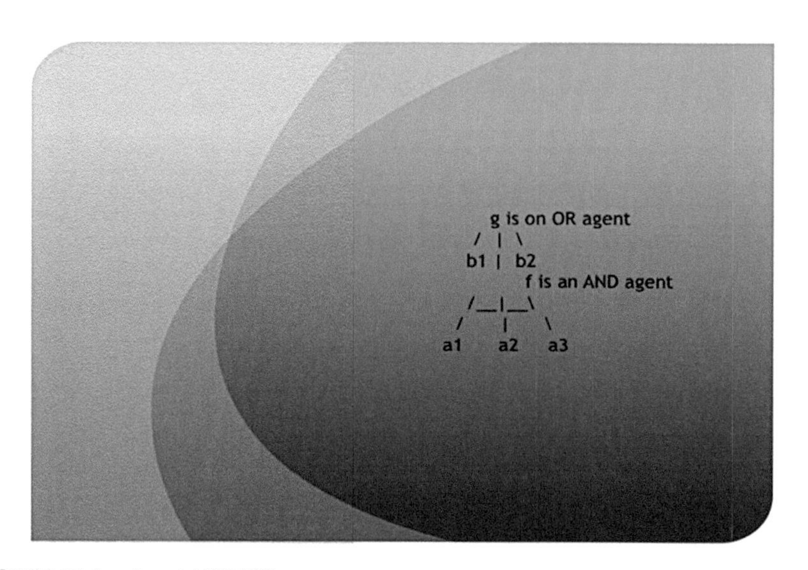

FIGURE 11.1 Agent AND/OR tree.

FIGURE 12.1 Morphs and trans-morphs.

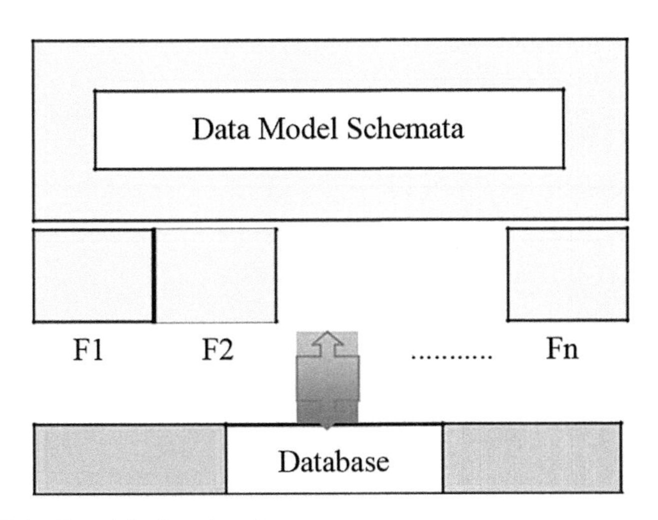

FIGURE 13.1 Keyed database interfaces.

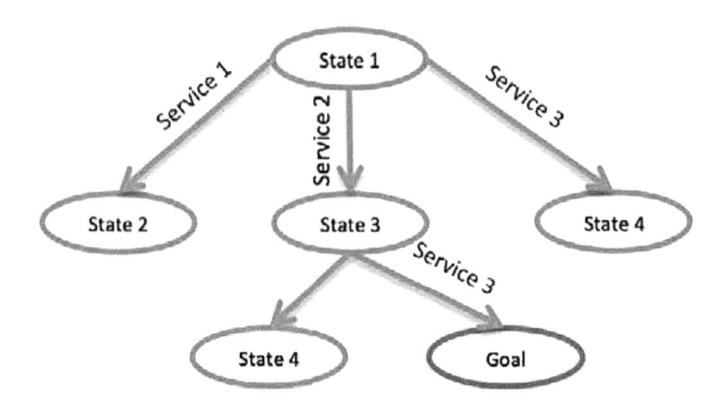

FIGURE 13.2 Decision Tree on a planning problem for service selection to reach a goal state.

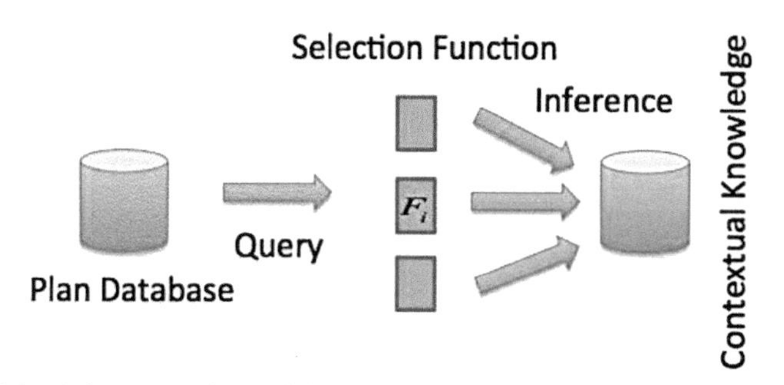

FIGURE 13.3 Inference engine and database inference.

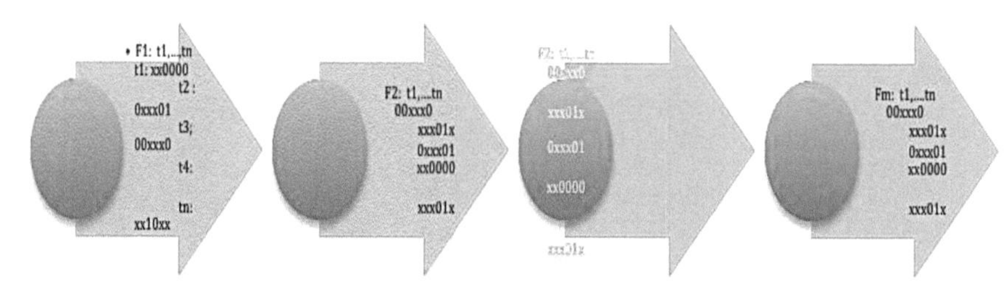

FIGURE 13.4 Sparse nonderministic products.

CHAPTER 8

SILICON VALLEY: TOO MUCH SUCCESS?[a]

HENRY ETZKOWITZ[1] and ANNIKA STEIBER[2]

[1]*International Triple Helix Institute, Silicon Valley, 1470 Sand Hill Road Suite 101, Palo Alto, CA 94304, USA*

[2]*Menlo College, College in Atherton, California, USA*

8.1 INTRODUCTION

A high-tech conurbation with an expansionary dynamic respects no bounds of nature, counter-culture, exurban or urban life. Starting from Santa Clara County on the Peninsula, Silicon Valley is expanding in all directions. Crossing the Santa Cruz Mountains to reach the Pacific coast, it is expanding into the city of San Jose as firms, like Google and Apple, outgrow the willingness of smaller cities such as Mountain View and Cupertino to accommodate their growth. Moving into and above Berkeley, it is spreading across counties formerly considered as part of the Bay Area, itself an expanding geographical classification. Even Oakland's downtown, where murals hid some empty storefronts, is experiencing signs of gentrification. Moving ever further east and south, Silicon Valley is expected to cross the mountains into the Central Valley where the University of California, Merced, a new campus, provides an anchor for future high-tech agglomeration in an agricultural region, much like the Valley itself 60 years ago. Indeed, San Joaquin County has pro-actively defined itself as "Greater Silicon Valley" as part of a concerted effort to promote the nascent trend.

[a]Presented at the session on "Science Scapes" Heidelberg Triple Helix Conference, 2016.

A new growth dynamic emanating from further expansion of mega-firms like Google, Apple and Facebook as well as start-ups converging from the rest of the world overlays the classic dynamic of university-based start-ups and local firm spinoffs. Indeed, the Stanford Research Park, whose early mission was to host firms that wished to remain close to their source, is virtually invisible in Silicon Valley, although it occupies a large tract of land on the east side of the Stanford campus. The physical format of low-lying buildings surrounded by green space became a model for the development of science parks in other locations where it was often presumed that the architectural format, in itself, was the attractor and generator of high tech development. Nevertheless, the Stanford Park is not seen as a significant factor in the development of Silicon Valley even though it served as the model for the contemporary science park.

Founded as an industrial park to attract manufacturing firms departing San Francisco and to raise money to support the development of Stanford, its founders soon realized that its potential resided in hosting firms emanating from Stanford that wished to stay close to their source for ease of continuing interaction. Today, the Park hosts the headquarters of the two descendant firms of Hewlett Packard, the Skype subsidiary of Microsoft, various law firms, the Start-X accelerator and other elements of the local innovation ecosystem. Silicon Valley's current employment growth is driven both by the location of branches of firms from Asian and Europe seeking to tap into the region's technology as well as the expansion of indigenous firms, both iconic and startups in long-standing and emerging technology fields. Ironically, quality of life is driven down for all but a super-elite as the result of an imbalance that emerges between private and public spheres. The traditional idea of management is to transform bad problems into good problems. 'Wicked problems,' have been defined as complex issues, the entanglement of multiple causations whose solution creates new innovation potential from the collaborative effort to meet their challenge (Oksanen & Hautamäki, 2015). This chapter discusses the sources of Silicon Valley's success and issues that have arisen due to "too much success, inducing, a "Katrina effect," similar to the long-term consequences of the hurricane for New Orleans.

8.2 EMPIRICAL RESEARCH: METHODOLOGY

Silicon Valley's "paradox of success" led us to define three main categories of research questions. The first category focused on how our interviewees defined 'Silicon Valley' and if 'Silicon Valley, according to them has changed over time. The second category of questions aimed to investigate to what extent Silicon Valley is 'place-bound' and if this is the case what makes it place-bound. The final category of questions tried to investigate if there is a 'primary engine for renewal' in Silicon Valley, and if so what are the sources for this and what risks are there that could "damage" this engine for renewal?

Twenty face-to-face "elite interviews" were carried out during the period from July to October 2015. Our aim was to both identify people with deep knowledge of the Valley but also cover perspectives from several parts of the ecosystem. For these reasons, we selected senior people from organizational bodies such as Silicon Valley Leadership Group, Bay Area Council, California Workforce Investment Board, Joint Venture Silicon Valley, and Collaborative Economics in Bay Area. In addition, we selected a high profile journalist who has followed the tech sector in the Valley for decades, two senior members of the VC community, two historians at Stanford University Library, people from the office of Economic Development in San Jose as well as several people representing the corporate view but also had in different ways reflected on Silicon Valley's trajectory.

The interview guide was semi-structured with open-ended questions, which provides us the opportunity to identify both similarities and differences in answers. Further, each interview was one hour long and was conducted either face2face or over Skype with the video camera. Each interview was documented and shared between the members of the research team. Data were coded by each one of the members of the research team and main findings were then jointly identified in frequent meetings during spring 2016 between members of the research team. Follow up questions identified were sent during spring 2016 to relevant interviewees and their answers were then integrated into the final documentation of key findings.

8.2.1 SILICON VALLEY'S "SECRET SAUCE"

Silicon Valley's extraordinary success came as a result of following the classic US "Endless Frontier" innovation model of concentrating resources and "picking winners" behind a laissez-faire façade. The original source of the Valley is a university with porous boundaries. The founding leadership, including Stanford University's President, David Starr Jordan, encouraged graduates to form technology firms in the late 19th century to electrify the region, utilizing existing technology. The next generation of Stanford faculty members, exemplified by Frederick Terman, together with their students, interacted closely with the next generation of firms, pursuing incremental innovation. In this era, the firms were often more technologically advanced than the university and aided its development.

The dynamic was set in motion, drawing technological demand into the university and sending research results out through cooperative relations with firms. Faculty were allowed and encouraged to have serious dual roles in firms. Technical industry existed in symbiosis with the university, indicated by a significant percentage of faculty recruited for impact and encouraged to continue extra-academic pursuits, to this date. A similar university-industry interaction dynamic occurred at MIT even earlier. This interactive dynamic is the source of new high tech conurbations and can be found in contemporary Pittsburgh in Carnegie Mellon University's attracting significant federal R&D funds, serving as the progenitor of that city's emerging robotics and AI industries.

The key intervening factor triple helix development in Silicon Valley was large-scale government funding of academic research, allowing a small-scale nascent process, exemplified by the founding of Hewlett Packard from a Stanford research project that had produced an innovative technology just prior to the World War II, to become an efficient breeder of startups in the post-war. Stanford drew government more tightly into its orbit during the early post-war by establishing Stanford Research Institute (SRI) dedicated to attracting funds, including projects beyond the interest and capacity of individual professors. Spun off from the university in the wake of the anti-Vietnam War protests, the Institute played a key role in transforming Stanford into a federally funded research university. Silicon Valley's growth dynamic, based upon silicon chips, was set in motion by government transistor procurement policy. Seeking to miniaturize battlefield communications equipment, the US Army drove a learning curve in transistor development that led to the development of the integrated circuit.

The chain-link innovation model, linking demand-side firm innovation to the supply-side academic invention, captured only part of this dynamic. The cluster of firms that emanated from this triple helix interaction acquired the label of Silicon Valley in 1971. According to several of our interviewees the label 'Silicon Valley' is still today associated with "...a flexible, forward-thinking economy...a unique ecosystem with an entrepreneurial spirit" and as "...an attractive place to start a company."

In succeeding decades the dynamic was replicated in other technology domains, supported by an increasingly complex set of supporting actors, including venture capital firms, technology transfer offices, and other boundary spanners. However, the most fundamental dynamic in the Valley emanated from the interaction between university and industry, among firms and between government and these more visible actors. Behind the two PhD students who met at Stanford's computer science department and became Google's founders was a Defense Advanced Research Project (DARPA) program that funded the research group that they were a part and posed the search problem that they solved.

It is a classic fallacy of "misplaced concreteness" that a Science Park with a set of buildings or a formal enclosed institutional format such as a Technopole can substitute for such an interactive dynamic. Unfortunately, this is the message that is most often taken away from Silicon Valley by visitors looking for a "quick fix" to achieve a knowledge-based conurbation without serious institutional restructuring, new institution formation, as well as long-term perspective and commitment. Firm growth support structure such as accelerators, providing mentoring and even financing are a significant advance beyond the brick and mortar Science Park model. Indeed, contemporary Science Parks have often repurposed themselves to act as accelerators as well as repositories for successful firms.

The key element of such accelerators, like Silicon Valley's Y-Combinator and StartX, is a training process through selection, insertion into a network of fellow startups, mentoring by experienced entrepreneurs and access to seed investment opportunities. The accelerator format rests upon an already developed high tech environment replete with a deep bench of angel investors, venture capital firms, potential startup collaborators that makes it possible for the accelerator supported firms to takeoff and flourish. That innovation ecosystem is itself is a second-order phenomenon, resting on the first-order dynamic of triple helix interactions among institutions with porous boundaries.

8.2.2 SILICON VALLEY'S "TRIPLE HELIX"

A Triple Helix with integrative boundary spaces (See Etzkowitz and Champenois, 2017) and institutional spheres that "take the role of the other" models a spiraling innovation process in which gaps may be filled by substitution of one actor by another. It is this latter capacity that makes the Triple Helix especially relevant to developing and declining industrial regions, alike. Indeed, the two prototypical US Triple Helix regions 'Silicon Valley' and Boston, in their early 20th century conditions, arose from collaboration and policy support.

An imbalanced triple helix, with the industry helix growing out of balance with government and university is a third order phenomenon, characterized by escalating inequality and a talent-gap. Interestingly, some of our interviewees viewed this innovation system as perilous; "Silicon Valley cares about everything inside the system but not what happens outside it. It is a navel-gazing economy…a parochial economy that feeds on itself and it lacks reflective capacity from a social perspective. These negative reflections of the phenomenon 'Silicon Valley' might be viewed as a negative side effect of a highly effective innovation system that primarily focuses on breaking new ground and creating the next unicorn, without taking into account the physical and social infrastructure required for its long-term support and renewal.

Of course, an imbalanced triple helix is a distant problem for regions struggling to take effective first steps to induce an innovation dynamic. When the 'cart is placed before the horse,' as when the Brazilian military regime constructed science parks in isolated suburban regions during the 1960's, little innovation activity occurred until a smaller scale model of incubators and entrepreneurial education within universities was adapted. At best, branches of existing firms and government laboratories may be attracted to a stand-alone Science Park. Some decades later when they close or downsize, their former employees who wish to stay in the area, may generate a startup dynamic as in Sophia Antipolis and Research Triangle. A more direct route is focused on facilitating university-industry interactions, especially creating an academic environment that recognizes it as a valued activity. The entrepreneurial university, holding a commitment to its region's development, with a significant number of faculty members who encourage their graduates to spin-off technology from their well-funded labs, and may hold dual roles in high tech firms themselves, are the core of a triple helix dynamic.

8.2.3 "TOO MUCH SUCCESS"

San Francisco, a traditional financial, manufacturing, and port city, as well as a tourist destination, cultural, and counter-cultural mecca are being swallowed by Silicon Valley.

According to all interviewees, Silicon Valley has expanded geographically and consists according to one of the interviewees today of "…a network of significant innovation nodes…there is now a Hardware node in the south, a Social Media node in San Francisco and Software oriented nodes everywhere." Another interviewee held that "Now there exist three epicenters in the Valley, East Bay, Santa Clara/San Mateo County, and San Francisco…"

Historically a working-class city with a small upper class, San Francisco is being transformed into an upper-middle-class city. Its working and middle classes are increasingly squeezed out due to escalating housing costs that are an unintended but entirely foreseeable consequence of an urbanization and rapid employment growth in high tech industries during the past decade. In 2011, firms like Twitter following the usual business tactic of threatening to leave the city, or not locate projected facilities, if their demands were not met, obtained tax breaks for a number of years. These were granted by the city administration on the condition that they locate their offices in a downscale area, the "Tenderloin," which the city wished to upgrade.

The influx is generated not only by firms locating in the city but also by employees of high tech firms on the suburban peninsula who prefer an urban lifestyle. Their employers, utilizing luxury bus coaches, to take them to and from work, have put on an ad-hoc inter-urban transportation system. Its highly visible presence, in contrast to the relative privacy in which residential succession takes place, has provided a focal point for anti-gentrification protests at the municipal bus stops that the private transportation system uses as its own. The busses are the most visible component of the "total institution" that these firms attempt to create in their office compounds, offering munificent snacks; gourmet free lunches and perks such as dry cleaning services to ease the burdens of everyday life, encouraging employees to focus on their work.

In just 5 years, Twitter and its peers attracted large numbers of employees who wanted to live in San Francisco. Rents have escalated, on the peninsula as well as in San Francisco and these areas have become unaffordable to many of its previous inhabitants who are constrained to move out. As one interviewee put it "Innovation is a hot topic here but we are in a danger to price ourselves out."

Even with rent controls in place, landlords may evict existing tenants and upgrade their premises to attract new tenants at double and triple rents. Provisions to have new upscale developments include a modicum of affordable units in exchange for increased density provides only a token solution. Older residents who can afford to remain oppose new high-rise structures, with, and without affordable housing elements, further exacerbating the housing crisis. Rents have recently dipped slightly, providing a breathing space to consider how to address the longer-term trend.

8.3 THE KATRINA EFFECT

The dislocation of people in New Orleans caused by the Katrina hurricane is happening not only in San Francisco, but in the whole Silicon Valley region from the influx of firms like Google, Facebook, Linked-In, and Twitter. According to Silicon Valley Index 2017, people are moving out of Silicon Valley nearly as quickly as they are moving in. Between July 2015 and July 2016, the region gained 22,500 foreign immigrants but lost 20,801 residents to other parts of California and the U.S.[1]

The unintended consequences of success are damaging the urban social fabric and causing attendant personal distress. What is happening in San Francisco as a persisting chronic condition is similar to what happened in a discrete acute way in New Orleans as a result of a natural disaster. People had to leave suddenly and move elsewhere. In succeeding years, some have moved back. Not surprisingly, people with greater resources have been more able to return to their native city. However, many with fewest resources who could ill afford to return have not come back. Also, not surprisingly, economic divides largely coincide with racial differences. In a sample of "... largely female African American poor people...," some leavers improved their housing and employment condition in the relocation process that was set in motion by the disaster but at the cost of loss of their connection to the culture of New Orleans (Waters, 2015). As a result, the city of New Orleans moved up a bit on the gentrification scale as its population shrank selectively.

The persisting negative consequence of the social change, even if positive in some respects, is "the Katrina Effect." Thus, impoverished former residents of New Orleans forced out by Katrina, improved their employment prospects and housing conditions. However, they suffered weakening or loss of social ties to family and friends by the physical distance imposed by relocation. Moreover, removal from the ambiance of their former familiar surroundings and even the loss of access to familiar food items within their cultural context caused further deprivation. The experience of material and psychic resources moving in opposite directions may also have a disconcerting, disorienting effect on those who experience it.

Whereas relative deprivation is the surplus of disesteem generated by comparison to peers who are otherwise equal (Merton, 1968); absolute deprivation arises from comparison with those who clearly advantaged on

[1]Presented by the Silicon Valley Institute for Regional Studies. Silicon Valley Indicators, www.siliconvalleyindicators.org/.

multiple dimensions. Absolute deprivation also generates its own surplus of psychological disesteem. However, the social distance between the lower and higher realms may also induce identification with the higher distant object as a substitute for attainment, or if social distance is so reduced as to allow contact, a breaking off of the relationship may ensue for fear of being dragged down, "a Rosalind effect," after the female character in *F Scott Fitzgerald's This Side of Paradise*, who breaks off her relationship with well-born, but increasingly impecunious, Amory Blaine.

An ever-present threat to the poor, eviction confronts middle-class San Franciscans with a new reality of downward mobility and exile. Exclusion is also experienced in its weaker form as lack of access to preferred housing location and type. A member of Palo Alto's Housing Commission, who recently resigned, illustrates the phenomenon. Both she and her husband, a software engineering, were relatively high paid but nevertheless could not afford to buy a house in Palo Alto. She announced in an interview that they were moving to Santa Cruz where they could find a house within their means. Ironically, the succession dynamic is replicated in Santa Cruz where it is reported that residents are moving out as housing costs grow beyond their means. The paradox of success creates a dynamic in which quality of life goes down except for the most highly successful who then drive up housing costs as they choose to live near where they work.

A social-ecological succession may be identified, similar to the natural one in which grasslands turnover into woods and back to grasslands again over a half-century or so. Artists, bohemians, and countercultural denizens in general, priced out of San Francisco by rising rents have been reported to be moving from San Francisco, reappearing in East Los Angeles, where they are seeding minority and working-class neighborhoods with a hip sensibility and attendant coffeehouse and restaurant businesses. It may be expected that more conventional young professionals will follow, attracted by the accouterments of a diverse urban neighborhood. Perhaps ironically, these were the very characteristics of the San Francisco scene that these artists had left, exemplifying another paradox of success.

There are inadequate means in place to deal with these issues. Joint Venture Silicon Valley (JVSV), is a public-private partnership that tracks trends on behalf of local governments and sponsors regular conferences to discuss issues. Business-led, it has convening but not governing power. There is no regional government in this region, with the exception of special purpose districts to deal with discreet phenomenon such as repairing the ecology of the San Francisco Bay or to provide community colleges through the Foothills District. Silicon Valley is a term popularized by a journalist in 1971, originally denoting the unique technological industry of the era, the microchip array of transistors, on which a succession of devices and industries have been built. The Silicon Valley label attracted international currency and took on a broader meaning representing the succession and intersection of technologies, the agglomeration of high tech, venture capital, universities, innovation, and entrepreneurial resources.

Silicon Valley's quintessential characteristic has been the informal technical community of crisscrossing firms, whose members exchanged information at local bars, after working hours in contrast to Boston's isolated large firms (Saxenian, 1994). Some of Silicon Valley's firms have recently set strict rules against discussing technical information with outsiders, compartmentalizing employees within the firm. Apple uses color-coded badges to maintain separation while Google has recently been faulted for falling behind in a key emerging technology, despite having greater expertise than its competitors. This expertise, however, was bottled up in discrete groups, working on separate smaller projects and products, rather than being brought together to achieve a larger goal an artificially intelligent personal assistant (O'Brien, 2016). Nevertheless, communication within Google, even across national boundaries is extensive (Steiber, 2014). The balance between secrecy and openness even in an era of 'open innovation' is still fraught.

The firm formation and growth engine that creates huge economic resources has attracted an influx of people globally, much as the gold rush of 1849, but the latter migration with its attendant industry of immigration lawyers is a longer-term phenomenon that shows no signs of abating. Whether keeping or discarding previous citizenship, they identify themselves as part of Silicon Valley. The migration includes firms as well as individuals and has generated a support structure to ease the entry and transition process. Many countries find it advantageous to establish "organizational beachheads," either their own incubators or space in existing facilities, to bring their startups to learn the methodology of high-tech growth.

In the opposite direction, leading Silicon Valley firms and universities search the world for talent and are usually successful in attracting it. A Stanford professor recently recounted how despite his position at a leading New York University (and an apartment in Greenwich Village) he accepted an offer to relocate. Silicon Valley is also a leading destination for successful startups from elsewhere in the world. Thus, "Not surprisingly the top 15 acquirers in the transatlantic ranking are all US companies. Even less surprisingly, 11 out of the top 15 are from the Silicon Valley" (Orizio, 2016). Open Austria, a three-person

Shop, sponsored by the Foreign Ministry and the Chamber of Commerce has opened an office in Galvanize, a private incubator in downtown San Francisco. Its mission is to assist startups from Austria visiting the Valley, seeking partners and resources and to keep an eye on trends in the startup and Silicon Valley technology and venture capital scenes. The Austrian Foreign Minister spoke at an opening breakfast event, attended by hundreds of Austria expatriates and friends, some of whom hold high positions in Google and other firms.

So is Silicon Valley place-bound or could it be replicated, and thereby threatened in its position as the leading innovation hub of the world? Our interviewees divided into two clear clusters. One group truly believed that Silicon Valley was place-bound "Terman tried to replicate the Valley to Korea but it didn't work. The concentration (in the Valley) is higher today than ever before and it is like a Black Hole with a great momentum." The other was more optimistic "...it is now replicated in Bangalore, Austin, and Europe." In the case the second cluster is right, Silicon Valley may experience less inflow of both great talent and companies in the future, assuming there is a finite limit that cannot be expanded upon. A third possible outcome is a dynamic interchange among multiple Silicon Valley's globally, expanding the talent pool by drawing in women and minorities now relatively excluded from participating in the Silicon Valley phenomenon.

8.3.1 UNINTENDED CONSEQUENCES OF "OVER SUCCESS"

Too much success is a broader phenomenon than Silicon Valley, with artistic impetuses as well, for example, in New York's Soho. The arrival of artists in a deindustrializing district, transforming factory buildings into lofts, followed by bars, galleries, restaurants; then attracts lawyers, stockbrokers, and other professionals who appreciate the bohemian ambiance generated. Loft prices are driven up and artistic community gradually departs as ever more upscale living places and businesses take root. The human ecological succession process has been made into a regional renewal tool: inviting artists to locate in order to jump-start the urban transition process. However, the transformative power of technology is arguably stronger than the artistic dynamic. Although, Walter Benjamin noted that art in the age of mechanical reproduction is duplicable; the hard copy distribution of Life Magazine images is modest in comparison to that of the Internet.

Silicon Valley is a global icon, a solution to the wicked problem of deindustrialization and underdevelopment, as well as a highly desired 'good problem'! Nevertheless, a 'corporate induced disaster' (Etzkowitz, 1984) is in the making if we consider the unwilling displacement of peoples as a deleterious consequence of innovation success. In Cambridge's Silicon Fen, green belt restrictions have pushed back against expansionist pressures, encouraging firms to relocate from the university town, on the one hand, while new high rise office blocks adjacent to the railway station provide some room for local growth. However, if office blocks are not complemented with housing, Cambridge will experience the same phenomenon as San Francisco, if it is not already.

A combination of public and private remedies may be suggested. On the one hand, revision of Proposition 13, the 1978 ballot initiative that reduced property taxes on individual houses and business property, is called for. Over time government lost more revenue and businesses gained much more than individual homeowners as business properties turn over much more slowly than houses and are thus less often subject to revaluation and potential rate rises. Companies should take responsibility for housing provision in conjunction with employment growth. Firms above a certain size might be required to provide a housing unit for each job created. In future development plans; corporate Campuses with housing as well as offices would be the mode. For example, Stockholm Kista Science Park where housing has been constructed could be the model for Stanford's science park.

Silicon Valley is the outcome of a nested "egg within an egg" model of porous university boundaries that encouraged a startup and spin-off dynamic and munificent government research funding that became the basis for "ecosystem" of intellectual property law firms, angel investors, venture capital firms; accelerators, incubators etc. A relatively a few decades ago, fruit orchards and a university were the highlights of the area that later became known as Silicon Valley. This quintessential high-tech conurbation, its label originating with Silicon chips, extends across an array of physical, software, and biologically based technologies, intersecting and hybridizing to create new industries and transform existing ones.

In Triple Helix and Innovation Studies, we usually inquire how to create a science and knowledge-based conurbation and focus our analysis on how to reach that objective. A variety of methodologies from Porter's diamond, cluster policy, Science Parks, and the European Union's smart specialization strategy have been invented to assist localities, regions, and nations in their quest to duplicate this success. The impetuses of declining industries and movement of high-paid jobs to low waged areas have lent urgency to this task. Other regions replication of Silicon Valley is a continuing challenge;

how can Silicon Valley restructure itself to respond to the Katrina effect is a new challenge.

The broader implications of the "Katrina effect" are exemplified by the Brexit vote in the UK and the Trump victory in the US: populations excluded from economic success in democratic countries will make themselves felt at the ballot box, even if the specific vote is not directly aligned with their dissatisfaction. An economy focused on an elite ("great financial payoffs" was by some of the interviewees viewed as the engine behind Silicon Valley's constant renewal), whether financial services or a high tech conurbation based upon the design but not the manufacture of devices is too narrow to provide sufficient economic opportunities for the majority of the population. This, in turn, will, according to several of our interviewees, become the greatest risk to Silicon Valley's engine of renewal; "Income disparities are more pronounced than ever, changing the character of our region and raising profound questions about community and cohesiveness."

An imbalance between public and private, with a preponderance of economic benefits flowing to a small elite rather than being spent on public goods, like infrastructure, education, R&D and healthcare, creating a broad range of economic opportunities and jobs, must, therefore, be redressed. The sources of Triple Helix innovation and entrepreneurship reside in both the public and private spheres. To over-emphasize the private at the expense of the public may produce a temporary advantage for a few but it will be at the expense of long-term sustainability.

8.4 CONCLUSIONS AND POLICY IMPLICATIONS

While other countries and regions outside of Silicon Valley are catching up in innovation capabilities, Silicon Valley's own overweening success has created an imbalance in the region that over time could damage the region's unique engine for renewal.

The academic sector is currently insufficient to supply talent and needs to be expanded dramatically, following the example of the Boston region (Etzkowitz, 2013). One temporary solution has been visa programs to import talent, but this talent will only be attracted to Silicon Valley as long as the region keep its unique status as a leading innovation hub in the world. From an era in which the academic sector was larger than the business community and actions needed to help create firms to employ graduates, the situation is now reversed. Firms' requirements for talent cannot be met locally. Drawing upon underutilized local talent will be an important part of the solution.

Firms are beginning to widen their recruitment efforts, based on demand but also as an effect of pressure from women and equality advocate allies as the percentage of women in tech is still low in the region.

Further, there is an imbalance in government, with an absence of regional government with the notable exception of a special district to save the bay from development and pollution. Industry-led efforts like Silicon Valley Leadership group has introduced a regional governance and have had some effect in supporting, e.g., regional mass transportation like the extension of the BART-system but little on the housing where the resistance to urban density is strong, especially among suburban homeowners who have paid high prices for their individual homes. The government was strongly involved in the origins and development of Silicon Valley, e.g., navy contracts to radio firm in the early 20th century, research contracts with SRI in early post-war, and semiconductor procurement from the 60's, but relatively absent in dealing with the consequences of success.

The future scenarios for the valley might be:

1. Continuation of current trends: the spread of Silicon Valley from 1–2 to 8–20 countries, with governance divided among numerous local entities, persisting inability to deal with broader issues;
2. Moderate change of course: partial repeal of Proposition 13, removing businesses from the reduced tax rates; thereby significantly increasing governmental abilities to fund traditional responsibilities like education that have severely declined in the wake of funding constraints; and
3. Radical course shift: the creation of City of Silicon Valley, including San Francisco, San Jose and the Peninsular in between, following the model of the creation of New York City in 1898, with elected mayor and council, with authority to address regional issues.

KEYWORDS

- **policy implications**
- **Silicon Valley**
- **Triple Helix Conference**

REFERENCES

Etzkowitz, H. (1984). Corporate Induced Disaster: Three Mile Island and the Delegitimization of Nuclear Power. Humanity and Society 8.3 1, Aug 1, 1984, http://search.proquest.com/openview/bbac9ee33ebb092490f17f05f015d19e/1?pq-origsite=gscholar.

Etzkowitz, H. (2013). Silicon Valley at Risk: The Sustainability of Innovative Region Social Science Information, *52*(3).

Etzkowitz, H., & Champenois, C. (2017). From the boundary line to boundary space: The creation of hybrid organizations as a Triple Helix micro-foundation. Technovation, forthcoming, and Etzkowitz and Zhou Triple Helix of University-Industry-Government Innovation and Entrepreneurship London: Routledge, 2017.

Merton, R. K. (1968). Social Theory and Social Structure. New York: Free Press.

O'Brien, C. (2016). Welcome to the new and expanded Silicon Valley Mercury News. August 13, 2016.

Oksanen, K., & Hautamäki, A. (2015). Sustainable Innovation: A Competitive Advantage for Innovation Ecosystems. *Technology Innovation Management Review, 5*(10), 24–30.

Orizio, S. (2016). "3 out of 4 startups are acquired by US companies." September 12. Mindthebridge.com Accessed 4 Oct. 2016.

Saxenian, A. (1994). Regional Advantage: Culture and Competition in Silicon Valley and Route 128. Cambridge: Harvard University Press.

Steiber, A. (2014). The Google Model: Management for Continuous Innovation in a Rapidly Changing World, Springer.

Waters, M. (2015). Disaster and Recovery: A Longitudinal Study of Hurricane Katrina Survivors. Stanford Sociology Colloquium, 19th February 2015. https://sociology.stanford.edu/events/sociology-department-colloquium-mary-c-waters last accessed 5 October 2016.

CHAPTER 9

ONTOLOGY-BASED APPROACH FOR REPRESENTATION OF SOFTWARE REQUIREMENTS

CHELLAMMAL SURIANARAYANAN,[1] DEEPA VIJAY,[2] and GOPINATH GANAPATHY[2]

[1]*Department of Computer Science, Bharathidasan University Constituent Arts and Science College, Tiruchirappalli, Tamil Nadu, India, E-mail: chelsrsd@rediffmail.com; chelsganesh@gmail.com*

[2]*School of Computer Science and Engineering, Bharathidasan University, Tiruchirappalli, Tamil Nadu, India, E-mail: deepa.vijay1@gmail.com; gganapathy@gmail.com*

ABSTRACT

A successful completion of a software project is purely depends on the better understanding of the requirements of the clients. Typically, a software team spends about 40 to 50% of their time on avoidable rework. As per the software industry standards, once the feature persists in production, cost of fixing the defects costs 100 times as high as it would have been during the requirements stage. In any software industry, rework of fixing the bugs is increasing due to poor understanding of requirements. Software requirements are mostly represented in natural language makes them prone to ambiguity and it also has its own limitation. The ambiguity of natural language leads to misunderstanding or poor understanding of requirements. There is a need to represent requirements in an unambiguous manner. In order to avoid ambiguity, an approach is proposed to represent the software requirements in semantic form using ontologies. The proposed approach is applied as a case study to the tourism-related application. The approach along with developed ontological concepts is presented. How it improves the understanding of requirements and thus minimizes the cost of rework are highlighted.

9.1 INTRODUCTION

The requirement gathering process is the first stage and an important stage in the software development process which has to be handled with care in order to achieve desired results. Requirements engineering plays a major role in software engineering discipline in achieving the desired outcome as quality products and services by meeting customer needs. The goal of requirement engineering is to achieve a common understanding of requirements among all stakeholders. Every Software Project Lifecycle begins from Software Requirement Elicitation, and requirements are the basis of every project defining the end user's expectation and the deliverables in order to fulfill goals of the particular project/product/services. Hence the success of any software project is purely based on good requirement specification.

Software Requirement Specification serves as a communicator and provides required information to various areas of the development cycle such as design, development, and testing. Drafting requirement specification is a challenging task. The team has to understand the purpose and need of the client, then write the Software Requirement Specification document to ensure that it works as a roadmap providing required information for the entire process in order to bring desired results.

In general, the Requirement Specification Documents are usually written in natural language by requirement team according specific standard such as IEEE Software Requirements Specification (SRS) 830. Requirements elicitation documents are created by its team members after gathering requirements from their clients, later it is mapped to concerned team members for further process. The development and other activities are based on these documents, and any error in the document always leads to bugs, increase of rework which in turn increases the cost of quality (Andrea & Vince, 2006; Jack, 1999) error in requirement specification brings huge expenses as the end products and services cannot be matched with customer needs. Though there are many scientific achievements and new technology in the software industry is in place, still few software products and services fail to produce desired results or fail to meet the end users requirements. This leads to dissatisfaction of clients. This can also cause a huge loss to the company sometimes even loss of business/clients.

Agile methodology is widely used in software industries in recent years due to its flexibility (Ming et al., 2004; Malik, 2011). The agile method relies on iterative development and face-to-face communication with the stakeholders. This approach of requirement gathering becomes very difficult for any project that follows a distribution model. In most cases, the software development activities take place in different geographic locations, wherein the clients are located at different geographic locations. Further, currently, all the requirements are represented in natural language. The requirements are created using their skills, knowledge, and level of understanding. So, it has its own limitation and often put teams in confusion. Because of the ambiguity of natural language, it leads to poor understanding of requirements, ambiguity in the specification of requirements and insufficient specification of requirements (Verónica et al., 2010). The need for alternate representation of requirements is exemplified by the following real scenario taken from our case study/project under study. As per the agile, the Requirements Elicitation is iterative. The requirements are represented in the form of user stories. User stories are not descriptive, and it is difficult to understand by the team members and hence they are unable to identify and derive multiple scenarios. Some of the issues faced in the project are given in the following real scenario.

9.1.1 REAL SCENARIO

9.1.1.1 REQUIREMENT OR USER STORY #1: IMPLEMENT HOTEL BOOKING FLOW IN AN APPLICATION

The above User Story was interpreted by the team members as: Perform Hotel Booking and the team has implemented this requirement. The functionality failed when a customer tried to perform a multi-room hotel booking. While analyzing the issue, the team understood this requirement as just a hotel booking and implemented for a single room as this is the default selection.

As requirement specifications are expressed completely by different means due to its flexibility the team members play a decisive role to ensure correctness and completeness of requirement. Team members toil by spending more time working on bugs rather than working on delivering new features. Hence, there is a real need in representing requirements without ambiguity for a quick and bug-free delivery. In this work, an alternate model based on semantics is proposed to eliminate ambiguity in the understanding of requirements.

The chapter is organized as different sections. Section 9.1 introduces the need for better representation of requirements. Section 9.2 gives the research works related to our theme. Section 9.3 presents the proposed model. In Section 9.4, a typical case study has been taken up to prove the proposed model. Section 9.5 concludes the chapter along with future work.

9.2 RELATED WORK

One of the ways to improve understanding of the requirements is representing the requirements in semantic form. In the early days, various knowledge representation languages like RDF, RDFS (Allemang & Hendler, 2008) were used. But they have some limitations. Expressivity of RDF and RDFS schema is very limited. Later OWL became popular (Smith et al., 2004). According to Gruber's definition (Gruber, 1993), "ontology is a formal explicit specification of a shared conceptualization." An ontology defines the common terms and concepts used to describe and represent an area of knowledge. Ontology can range in expressivity from taxonomy to thesaurus to a conceptual model (Ontology Transformation, 2009).

Though the natural language was well accepted and widely used for requirement specification it had many drawbacks. To overcome these issues, formal specification languages like Z and B (Tulika & Saurabh, 2015) were used. But the issues with the formal languages like Z and B, it is difficult to understand and implement as it is represented in mathematical form. Hence Ontology seems to be well suited for an evolutionary approach to the specification of requirements and domain knowledge. Moreover, ontologies can be used to support requirements management and traceability which helps the software engineers to better understand the relations and dependencies among various software artifacts (Diego et al., 2016). In the research work (Yuan & Zhang, 2015), an approach of interactive requirements elicitation for the production of customized software systems has been presented. This approach constructs an ontology model to represent software assets and to extract instances of the ontology model as abstract requirement models for specific application domains.

Research works such as Avdeenko & Pustovalova (2016) and Katja et al. (2012) employ ontology to create a system of requirements, satisfying the whole set of properties: correctness, completeness, consistency, unambiguity, traceability. In contrast to these approaches, the proposed research work presents an approach which develops ontological concepts to capture the semantics of requirement feature. This results in better understanding of feature which reduces the cost of rework and improves quality.

9.3 PROPOSED APPROACH

In this work, an ontology-based conceptual model is proposed to represent the requirements in semantic form using Web Ontology Language (OWL). Few requirements have been selected from an application and a case study is designed for developing a conceptual model. The approach is proposed with special emphasis to an agile method of developing software projects. Because agile method accommodates volatile requirements, focuses on collaboration between developers and customers, and supports early product delivery. Two of the most significant characteristics of the agile approaches is that they can handle unstable requirements throughout the development lifecycle and they deliver products in shorter timeframes when compared with traditional development methods. The model consists of two steps, namely, Step 1: develop micro-level ontologies; and Step 2: develop concrete scenarios.

9.3.1 STEP 1: DEVELOP MICRO-LEVEL ONTOLOGIES

For each requirement, a 'micro-level' ontology has been developed in order to annotate the semantics which avoids ambiguity in requirements and helps in deriving multiple scenarios. User stories are not descriptive (See User Story #1 in Section 9.1).

It is difficult to understand by the team members and hence they are unable to identify and derive multiple scenarios. So, the core terms of requirement/ user story are enhanced with explicit semantics. Meaningful concepts as per client's requirement are developed as ontologies using tools such as Protégé. As ontological concepts are developed for each requirement feature or user story we call them as micro level ontologies.

9.3.2 STEP 2: DERIVE CONCRETE SCENARIOS

The micro level ontologies are accessed using tools such as Jena API. The core terms are mapped to concepts of the developed ontologies. This semantic tagging removes the ambiguity in a misunderstanding of requirements. This facilitates the derivation of multiple scenarios with possible semantics corresponding to a core feature. The derived scenarios are refined with customer feedback in order to generate concrete scenarios. The two steps of the model are illustrated in Figure 9.1.

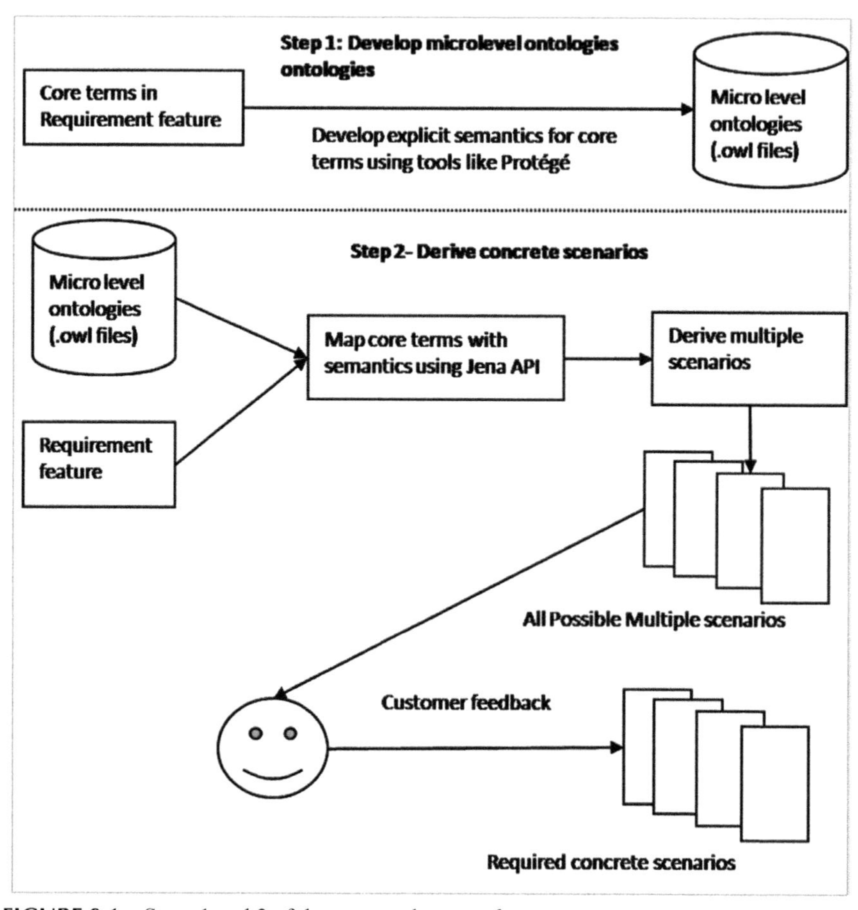

FIGURE 9.1 Steps 1 and 2 of the proposed approach.

The overall concept of the approach can be realized as given in Figure 9.2. Different micro ontologies are first constructed and they are used during requirements envisioning process. Ontology acts as a baseline document for detailed analysis of the requirements. Developed ontologies can be reasoned using various APIs like JENA.

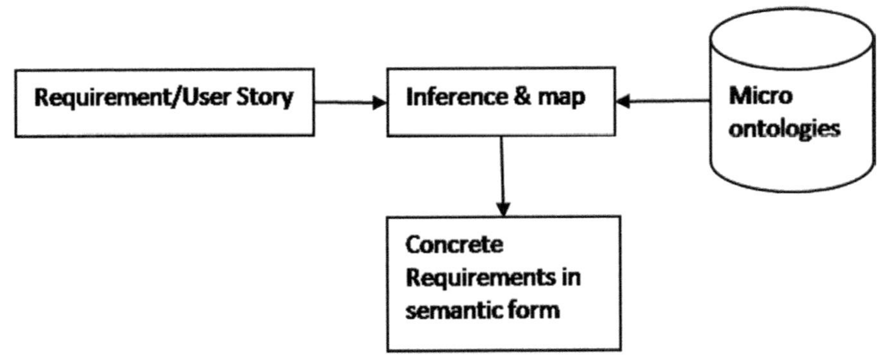

FIGURE 9.2 Realization of the proposed approach.

9.4 CASE STUDY

The plan of the research work is to develop micro-level ontologies which correspond to atomic features. Then, these micro ontologies are combined into application level ontology which is capable of representing all the features in an application. Finally, all application ontologies may be combined to serve as a domain level ontology. An application in the tourism domain is considered as the typical case for the study. This application is an online travel agency serving the customers worldwide. The case under study has followed an agile scrum framework for over 3 years. The project followed Onsite-Offshore (Distributed) model. Three peoples including Product Owner sit at Onsite and 15 peoples at offshore. The team followed the agile scrum framework (Astha & Divya, 2014) as described in Figure 9.3.

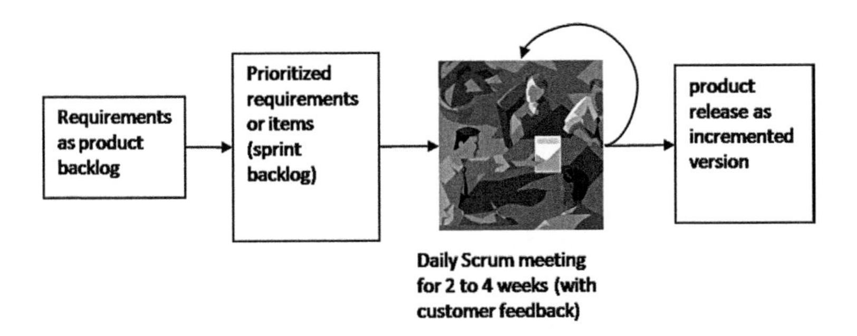

FIGURE 9.3 (See color insert.) Scrum sprint cycle.

Sprint cycle has lasted for 2 weeks. All the requirements are placed in product backlog and maintained by the product owner. Sprint planning meeting is scheduled before every sprint. Product owner prioritizes the sprint items and the offshore team provides estimates for each user stories. Once the sprint backlog is ready, the team starts working on sprint items. Envisioning, design, and development and testing activities were carried out throughout the sprint. Every day, daily scrum meeting was scheduled between onsite and offshore team members through telephone and discussed what has been done, what will be done, and any blockers. This meeting lasts for 15 minutes to 30 minutes. At the end of each iteration, sprint demo is shown to the customers for their feedback.

This case has been analyzed for the factors and root causes that lead to defects that occurred during the development of the travel module in the year 2012. The root cause analysis has been analyzed in our previous work (Gopinath & Deepa, 2013). Percentage of defect count due to poor understanding of requirements are analyzed as given in Table 9.1 (Gopinath & Deepa, 2013).

TABLE 9.1 Defect Count and Root Cause Reasons

#	No. of Defects	Where/why the defects are missed	% of contribution	% of defect count and primary cause
1	36	Missed in Envisioning	25.53	60.29%of defect count is primarily due to poor understanding of requirements
2	5	Missing Component Test Coverage	3.55	
3	44	Missing Unit Test Coverage	31.21	
4	3	Missed in Integration Test Coverage	2.13	Remaining 39.71% defect count is due to other reasons
5	2	Missing Usability Test Coverage	1.42	
6	6	Missed due to Environment Limitation	4.26	
7	45	Missed due to some other reasons	31.91	

From Table 9.1, it is clear that 25.53% of the defects were leaked due to Missed in Envisioning (Escaped in Envisioning activity) and hence 31.21% of the defects were leaked due to Missing Unit Test Coverage as well as 3.55% of the defects leaked due to missing component test coverage. So, the first three primary root causes are missing defects in envisioning and unit testing and component test). It is has been strongly realized that the reason for the defects is primarily due to poor understanding of requirements. So, this kind of product development has been taken as the typical case for our present study. In order to alleviate the problems associated with an understanding of the requirements, the proposed approach has been employed in this case as given below.

At first micro-level ontologies have been developed for different requirement features for the case under study. Let us consider an application specific requirement which is taken from the Travel domain project.

Requirement or User Story: Implement Hotel booking flow in an application

Concept -> Hotel

In the above requirement, 'Hotel' is a generic word. It has multiple scenarios which can be understood by a domain expert. It would be very difficult to understand and derive multiple scenarios for new resources without any Travel domain knowledge. The above user story can be represented in semantic form as below:

Implement Hotel (MyHotel.owl) booking flow in an application has been developed as a micro level ontology as given in Figure 9.4.

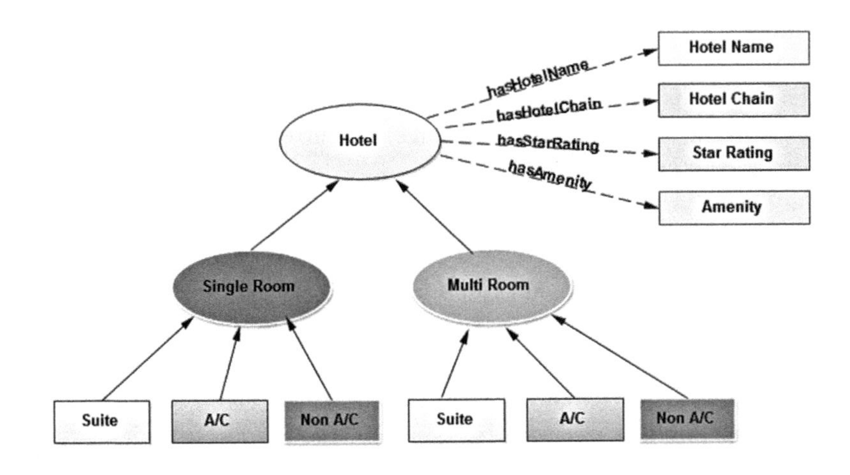

FIGURE 9.4 MyHotel.owl.

The semantic mapping in Figure 9.3 gives a better understanding of the user story (requirement). This requirement mapping is very easy to understand and execute. Even the person who is ignorant of domain and application knowledge will be able to understand and derive multiple scenarios.

The following scenarios are derived from the Figure 9.4:

- Perform Hotel Booking for Single room suite;
- Perform Hotel Booking for Single room A/C;
- Perform Hotel Booking for Single room Non-A/C;
- Perform Hotel Booking for Multi-room suite;
- Perform Hotel Booking for Multi-room A/C;
- Perform Hotel Booking for Multi-room Non-A/C.

User Story #1

Summary: Add option to create generic business pricing profile in Business Tool.

Description: We need to add an option to create a business pricing profile in Business Tool. This will be added in the page to create a business pricing profile, we need to display an option "Is Profile reusable," checking this should make the profile a reusable one.

When the option to make the profile generic is selected, distributor section should be disabled.

User Story #2

Summary: Add multi-language and currency options in Business Tool for Creating business pricing profiles.

Description: We need to add language and currency in the Business Tool to create business pricing profile.

The User Story 1 and 2 are mapped to the semantic representation as given in Figure 9.5.

User Story #3

Summary: User should be able to perform a search based on Hotel Chain Code.

Description: User should be able to perform a search based on Hotel Chain Code.

User Story #4

Summary: User should be able to perform a search based on Hotel Name.

Description: User should be able to perform a search based on Hotel Name.

User Story #5

Summary: User should be able to perform a search based on Hotel Star Rating.

Description: User should be able to perform a search based on Hotel Star Rating.

User Story #6

Summary: User should be able to perform a search based on Hotel Amenity

Description: User should be able to perform a search based on Hotel Amenity

User stories 3 to 6 are mapped to the existing ontology (MyHotel.owl) as represented in Figure 9.4.

User Story #7

Summary: Modify Business Pricing Report.

Description: As an Admin user, one should be able to retrieve business-pricing report with all details.

User Story #7 is implemented as given in Figure 9.6.

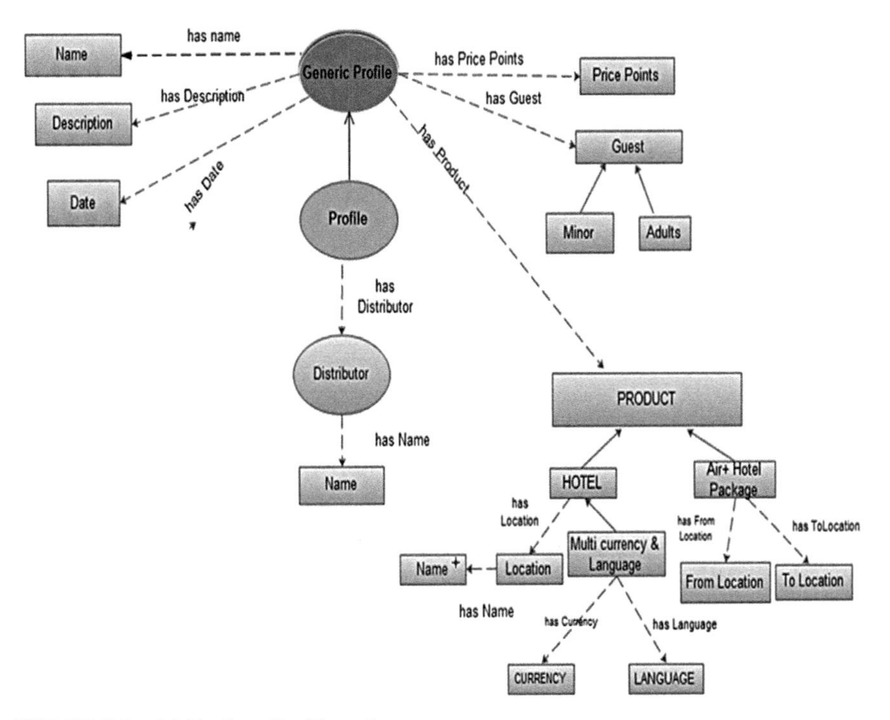

FIGURE 9.5 MyBusinessProfile.owl.

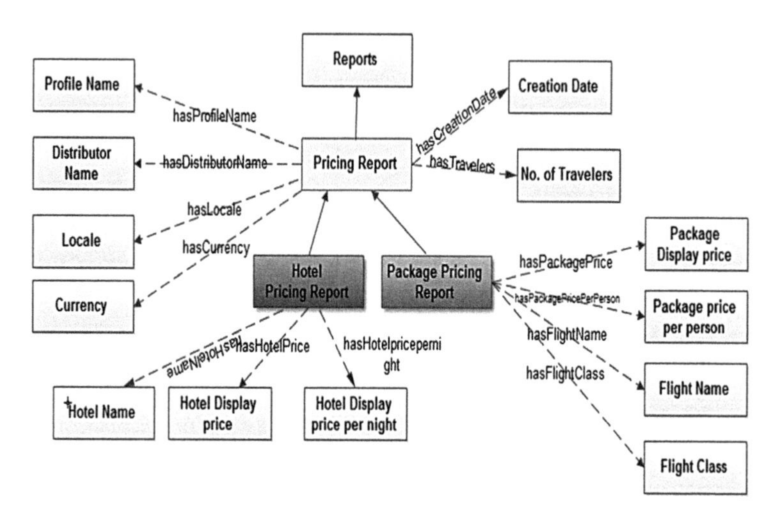

FIGURE 9.6 MyBusinessReport.owl.

User Story #8

Summary: Add Search Widget to Business Homepage

Description: Search Widget should be available on the Business Homepage.

User Story #8 is implemented as given in Figure 9.7.

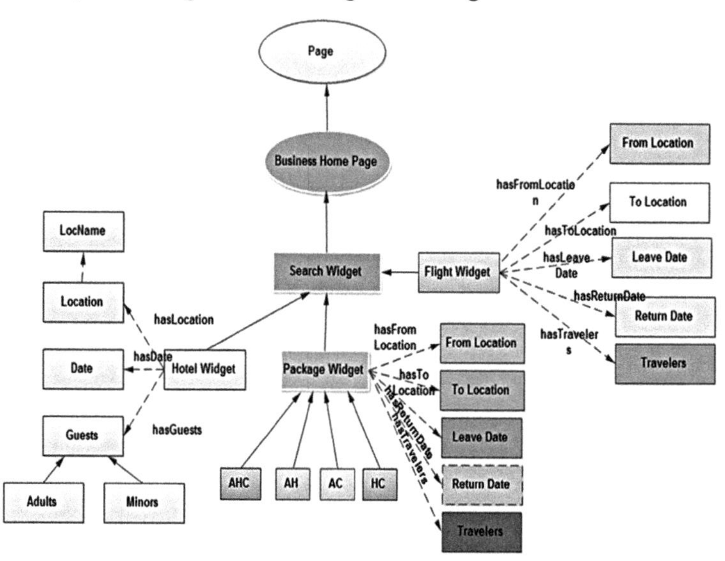

FIGURE 9.7 MyBusinessHomePage.owl

From the implementation of envisioning stage with the above set of micro-level ontologies, the bugs in the requirement envisioning processing have been reduced significantly around 88.89%. Out of 9 user stories taken up for ontological representation one user story namely, *Hotel price should be rounded off in an application alone* could not be represented in the ontology as the requirement is syntactically incomplete. So, it is recommended that the requirements have to be verified for its completeness before implementing semantics. Further, Semantic representation of the behavior for the function "rounded off" is difficult at this stage of research.

9.5 CONCLUSION

In this research work, an approach has been presented to represent requirement features of software projects in order to avoid ambiguity in the understanding of requirements. In the proposed approach, micro level ontologies corresponding to different atomic requirement features have been developed. From these ontologies, the development team maps all possible scenarios for a given user story. From the all possible scenarios, required scenarios have been identified as concrete scenarios after discussion with customers and from their feedback.

When requirements are specified with our approach, it is found that better understanding of requirements reduces the number of bugs and hence reduces the amount of rework involved in fixing the bugs. This significantly improves the cost of software quality. Further, from the case study, it is recommended that the semantic representation of ontological concepts and semantics should be followed after the verification of completeness of requirements in its syntactic format.

KEYWORDS

- **cost of software quality**
- **ontology-based requirements representation**
- **software requirements specification**
- **tourism application**

REFERENCES

Allemang, D., & Hendler, J. A., (2008). *Semantic Web for the Working Ontologist*. Modeling in RDF, RDFS, and OWL. Elsevier, Amsterdam.

Andrea, S., & Vince, T., (2006). "A review of research on the cost of quality models and best practices," *International Journal of Quality & Reliability Management, 23*(6), 647–669.

Astha, S., & Divya, G., (2014). "Scrum: An agile method." *International Journal of Engineering Technology, Management, and Applied Sciences, 2*(6), 182–190.

Avdeenko, T. V., & Pustovalova, N. V., (2016). "The ontology-based approach to support the requirements engineering process." *13th International Scientific-Technical Conference on Actual Problems of Electronic Instrument Engineering (APEIE),* Novosibirsk, pp. 513–518

Diego, D., & Jéssyka, V., (2016). Ig Ibert Bittencourt, "Applications of ontologies in requirements engineering: A systematic review of the literature," *Requirements Engineering* (Vol. 21, No. 4, pp. 405–437). Springer.

Gopinath, G., & Deepa, V., (2013). "Empirical case study of agile scrum process." In: The *Proceeding of the Indian Conference on Research Ideas in Software Engineering and Security RISES* –19/08/2013–20/08/, Madurai.

Gruber, T. R., (1993). "A translation approach to portable ontology specification." *Knowledge Acquisition, 5*, 199–220.

Jack, C., (1999). *Principles of Quality Costs: Principles Implementation and Use (3rd edn.)*. ASQC Press, Milwaukee, pp. 90–102.

Justas Trinkunas and Olegas Vasilecas (2009). *Ontology Transformation: From Requirements to Conceptual Model*, Scientific Papers, University of Latvia. Vol. 751.

Katja, S., Yuting, Z., Jeff, Z. P., & Uwe, A, (2012), "Measure Software Requirement Specifications by Ontology Reasoning," Conference: Proceedings of the 8th International Workshop on Semantic Web Enabled Software Engineering (SWESE 2012).

Malik, F. S., (2011). "An agile software development framework." In: *Proceedings of International Journal on Software Engineering (IJSE)* (Vol. 2, No. 5).

Ming, H., June, V., Liming, Z., & Muhammad, A. B, (2004), "Software quality and agile methods." In: *Proceedings of Annual International Computer Software and Applications Conference (COMPSAC'04).*

Smith, M., Welty, C., & McGuiness, D., (2004). Owl web ontology language guide. *Recommendation W3C, 2*(1).

Tulika, P., & Saurabh, S., (2015). "Comparative analysis of formal specification languages Z, VDM, and B." *International Journal of Current Engineering and Technology, 5*(3).

Verónica, C., Luciana, B., Ma. Laura, C., & Ma. R. G., (2010). " The use of ontologies in requirements engineering." *Global Journal of Research in Engineering [Online], 10*(6), 2–8.

Yuan, X., & Zhang, X., (2015). "An ontology-based requirement modeling for interactive software customization," *2015 IEEE International Model-Driven Requirements Engineering Workshop (MoDRE)*, Ottawa, ON, pp. 1–10.

CHAPTER 10

COMPETITIVE MODEL BUSINESS ANALYTICS AND ERP OPEN LOOP OPTIMIZATION

CYRUS F. NOURANI

AKDAFW GmbH Berlin, Germany

ABSTRACT

New optimality principles are put forth based on competitive model business planning. A generalized MinMax local optimum dynamic programming algorithm is presented and applied to business model computing where predictive techniques can determine local optima. Based on a systems model an enterprise is not viewed as the sum of its component elements, but the product of their interactions. The chapter starts by introducing a systems approach to business modeling. A competitive business modeling technique, based on the author's planning techniques is applied. Systemic decisions are based on common organizational goals, and as such business planning and resource assignments should strive to satisfy higher organizational goals. It is critical to understand how different decisions affect and influence one another. Here, a business planning example is presented where systems thinking technique, using causal loops, are applied to complex management decisions. Predictive modeling specifics are briefed. A preliminary optimal game modeling technique is presented in brief. Conducting gap and risk analysis can assist with this process. Example application areas to e-commerce with management simulation models are examined.

10.1 INTRODUCTION

The chapter starts with an introduction to systems thinking and its applications to business modeling. Systems thinking techniques and modeling principles are presented first. Then, business planning with goals and open loop control based on prediction is introduced. An example of systems modeling using causal loops is presented. Competitive business modeling (Nourani, 2002) is then introduced and game tree competitive learning is discussed with applications to systems, goals, and models. Practical planning systems are designed by modeling the business with tiers. A new 'competitive model dynamic programming technique and algorithm is introduced where the stages are optimized based on predictive models. Lack of an all-encompassing and holistic approach is one of the key contributors to organizational downfall as this can lead to over- fragmentation, cross-purposes, loss of resources, and internal competition within organizations. Systems thinking is a paradigm and set of methodologies that allow individuals and organizations to adjust their mental models to a holistic view and systemic nature of relationships. Enterprise planning is an integrated planning concept and associated models aiming to create a seamless approach to organizational coordination and cooperation.

Hence, Enterprise Research Planning (ERP) is rooted in the systemic view of the organization. This chapter discusses ERP in the context of systems thinking and modeling. Corporations have to identify and specify their business processes before they implement an ERP system. That is best accomplished by having a specific planning system to realize goals. ERP and Experience Management (EM) are to be applied to elect supply chain policies, which can in part specify how the business is to operate. Applying resource planning processes business planning models could be developed. ERP tactics might apply enterprise systems that can plan for critical resources up to sales and delivery. The planning process and ERP implement schedules and raise critical success factor problems to appropriate senior management. Conducting gap and risk analysis between functionality that will be implemented based on systems capability is an important prerequisite for a successful design for systems.

The ERP models presented here including multiplayer games draw from the author's projects. Our planning systems are designed by modeling with information, rules, goals, strategies, and knowledge bases. Finally, ERP is presented as a set of project tasks to satisfy business goals. Below is a schema of the business planning model. Agent games applications to economics date to author's (Nourani, 1999). Artificial adaptive agents (Bullard & Duffy, 1998), not in the exact sense as here, were applied to economic theories ever since (Holland & Miller, 1991; Miller et al., 2002). More recent applications with agent computing have appeared in Nourani and Maani (2003); Nourani

(2002) and at least a published book (Parsons et al., 2002). The chapter's organization is as follows: Section 10.1.1 starts with a business planning example, followed onto Section 10.2 on systems thinking, business modeling, goals, planning, prediction, and open loop control are presented. Specific predictive techniques are outlined with examples. Hypotheses formation is based on data discovery. Data models and discovery techniques are briefed in Section 10.5.2. Dynamic programming (Bellman, 1975) and goal achievements are introduced with competitive models applications to ERP following an example area. Game trees and competitive model planning with agent game trees is presented. Optimal games, plans, and uncertainty are examined with predictive model computing techniques, followed by introducing the plan optimality competitive model algorithm. Specific examples applications are presented to select business models. The techniques are further applied to e-commerce in light of "management flight simulator" techniques.

To compare with our techniques a basic book on the areas might be Fleisher, C.S. and B. Bensoussan, where Competitive intelligence practitioners, practicing managers and business decision-makers, are addressed on how to treat Business and Competitive Analysis from transforming raw data into compelling, actionable business recommendations. What is called for is guidance on the analysis process, including defining problems, avoiding analytical pitfalls, choosing tools, and communicating results? Enterprise analysis techniques, Product Line, and Win/Loss Analyses, Strategic Relationships, critical success factors, driving Forces, Technology Forecasting, Gaming, Event/Timeline, Indications are to be examined in competitive business planning. The basics that our techniques consider are Business Model Analysis, SERVO Analysis, for example, optimality parameters, Supply Chain Management, Product Line Analysis, Win/Loss Analysis, Event, and Timeline Analysis, Technology Forecasting, and games applications.

10.1.1 BUSINESS PLANNING EXAMPLE

The diagram (see Figure 10.1) is an overview of a business planning example carried onto the chapter. The arrows indicate feedback loops and lines denote forward predictive open control loops. ERP involves games (Nourani & Maani, 2003), open, and closed control loops, AND/OR causal trees, and splitting agent decision (Nourani, 1994; Nourani, 1996). Using these tools, management processes can be represented by system models. A business plan has a mission statement, technical design, required resources, a specific product, and a market forecast. To realize the goals, enterprise resource planning (ERP), predictive modeling, risk analysis, and supply chain considerations are to follow (Figure 10.1).

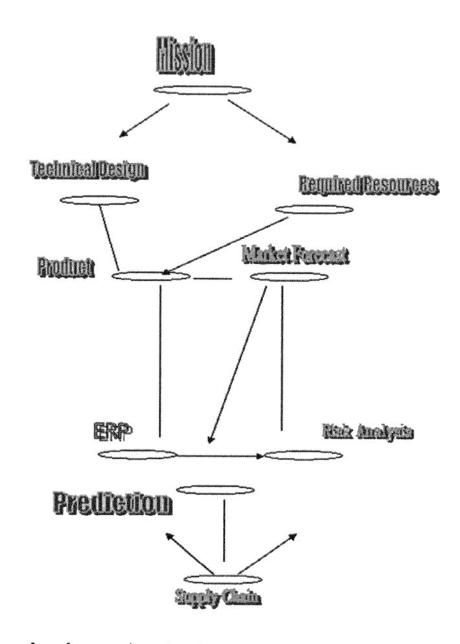

FIGURE 10.1 **(See color insert.)** A business planning example.

10.2 SYSTEMS MODELING

Systems thinking (ST) is a discipline for understanding complexity. This complexity underlies business, economic, scientific, and social systems. ST has three dimensions: paradigm, language, and methodology. These dimensions are briefed below with further specifics that are applied in Nourani & Maani (2003) as follows:

Paradigm: ST is a way of thinking about the world and relationships. This paradigm describes the dynamic relationships that influence the behavior of systems (wholes). This consists of three types of thinking:

1. Dynamic – recognizing that the world is not static and things change constantly;

2. Operational Thinking – understanding the 'physics' of operations and how things really work and interact; and
3. Closed-Loop Thinking – recognizing that cause and effect are not linear and often the end (effect) can influence the means (cause).

Language: As a language, ST provides a tool for understanding the complexity and dynamic decision-making. ST language (Anderson & Johnson, 1997).

Methodology: ST has a set of modeling technologies (Anderson & Johnson, 1997; Kauffman, 1980; Maani & Cavana, 2000). These tools can be used to understand, measure, and predict the behavior of systems, as well as to facilitate and accelerate group learning. These tools are Stock and Flow Diagrams, Microworlds (Computer Simulation), and Learning Laboratory.

10.3 PLANNING

ERP and EM are to be applied to elect supply chain policies, which can in part specify how the business is to operate. Applying resource-planning processes appropriate business planning models could be developed. Applying ERP to the planning process, we might develop tactical planning models that plan critical resources up to sales and delivery. Planning and tasking require the definition of their respective policies and processes; and the analyses of supply chain parameters. Dynamic thinking, that the business world is not static, is a fundamental premise to ERP.

Operational thinking is applied when planning for decisions based on how things really work and interact. Closed-loop thinking can be applied to control decision trees on ERP when feedback is necessary to carry on new plans or to set new business goals. Experience Management (EM) at times can assist in obtaining closed loop control or be applied to determine the scope of the project to fit within the resources available (including budget) and the time required and to assign responsibility.

ERP has to select the operational planning teams that will manage the process the following tasks: feasible production schedules, mapping the operational planning process to develop scheduling models for production and plan inventories, developing plans for building a data warehouse, Knowledge Management (KM), and interfaces to legacy systems. ERP involves implementation teams consisting of business representatives, functional representatives, IT personnel, and system knowledgeable people.

The planning and ERP team implement schedules and raise performance problems to appropriate senior management. Gap and risk analysis can be conducted to determine system capability. ERP and EM can be applied to determine flexibility factors to modify implementation plans, as specific functionality domains become critical.

10.3.1 GOALS, PREDICTIONS, AND OPEN LOOP CONTROL

Patterns, schemas, and viewpoints are the 'micros' to aggregate information onto the data and knowledge bases, where masses of data and their relationships and representations are stored respectively. Forward chaining is a goal satisfaction technique where inference rules are activated by data patterns, to sequentially get to a goal by applying the inference rules. The current pertinent rules are available at an 'agenda' store. The rules carried out will modify the database. Backward chaining is an alternative based on an opportunistic response to changing information. It starts with the goal and looks for available premises that might be satisfied to have gotten there. Goals are objects for which there is automatic goal generation of missing data at the goal by recursion backward chaining on the missing objects as sub-goals. Data unavailability implies a search for new goal discovery.

A basis to model discovery and prediction planning is presented in Nourani (2002) and Bellman (1975) and is briefed here. Minimal prediction is an artificial intelligence technique defined since the author's model-theoretic planning project. It is a cumulative nonmonotonic approximation attained with completing model diagrams on what might be true in a model or knowledge base. The prediction techniques are the basis for 'Open Loop' control explained earlier (Nourani, 2002).

10.3.2 DYNAMIC PROGRAMMING AND GOALS

Practical planning systems are designed by modeling the business with tiers (Nourani, 2005; Nourani, 2004). Patterns, schemas, AI frames and viewpoints are the micro to aggregate glimpses onto the database and knowledge bases were masses of data and their relationships-representations, respectively, are stored. Schemas and frames are what might be defined with objects (Nourani, 1997), the object classes, the object class inheritances, user-defined inheritance relations, and specific restrictions on the object, class, or frame slot types and behaviors. A schema could be intelligent forecasting, IS-A stock forecasting technique, portfolios stock, bonds, a corporate asset as well as Management Science techniques. Schemas allow brief descriptions on business object surface properties with which high-level inference and reasoning with incomplete knowledge can be carried out by defining information and relationships amongst objects. Relationships: Visual Objects A and B have a mutual agent visual message correspondence. Looking for patterns is a way some practical AI is carried to recognize important features, situations, and applicable rules. In this sense data patterns are akin to leaves on computing trees.

A scheme might be:

- Intelligent Forecasting;
- IS-A Stock Forecasting Technique;
- Portfolios Stock, Bonds, Corporate Assets;
- Member Management Science Techniques.

Dynamic programming is as follows: Break up a complex decision problem into a Sequence of smaller decision subproblems. Stages: one solves decision problems one “stage” at a time. Stages often can be thought of as “time” in most instances. We shall present specific dynamic programming techniques and applications in the following sections.

10.4 ERP AND COMPETITIVE MODELS AND GOALS

Competitive Models might be Micromanaged to access competitive goal satisfaction advantages. An example of competitive models will be presented via the transactional business model in the following section. Alternate models can be designed based on where assets, resources, and responsibility are assigned; how control and coordination are distributed; and where the plan goals are set. Applying System as Cause – (assumptions, beliefs, values, etc.) and System vs. Symptom principles and quality management where root causes of a ‘problem’ should be identified (i.e., Cause, and Effect Diagrams) is a precondition for finding lasting solutions. Time-space dynamics, short vs. long term and soft indicators are crucial to competitive modeling.

The qualitative systems modeling example can be extended to quantitative or dynamic system modeling. However, this topic is beyond the scope of this chapter. With dynamic modeling, ‘live’ simulations of real-life situations can be developed and different scenarios can be tested under a variety of assumptions and the results can be quantitatively assessed. Dynamics modeling takes into account the inherent uncertainty of the real world and allows different scenarios to be considered concurrently. Situation-based models (known as Management Flight Simulators) are often used in a 'learning laboratory' environment where management teams 'practice' hypothesized game plans and learn from experimentation and error. 'Soft indicators' and 'short and long-term' principles are critical in competitive learning situations.

Example: A business manager has 6 multitalented players, designed with personality codes indicated with codes on the following balls. The plan is to accomplish 5 tasks with persons with matching personality codes to the task, constituting a team. Team competitive models can be generated by comparing teams on specific assignments based on the task area strength. The optimality principle might be to accomplish the goal with as few a team grouping as possible, thereby minimizing costs (Figure 10.2).

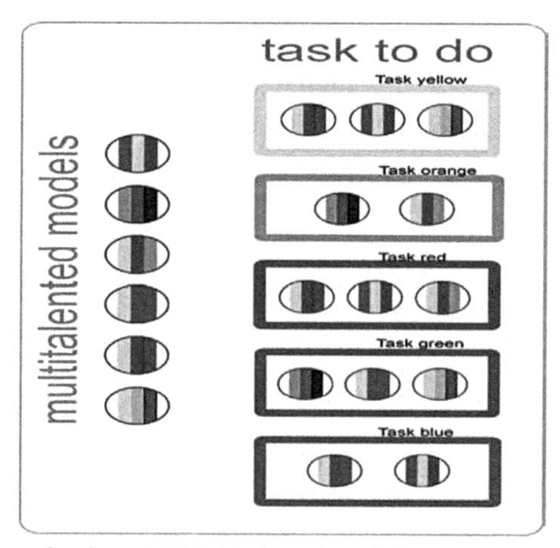

FIGURE 10.2 **(See color insert.)** Multitalented models.

10.4.1 GAMES

Game theory, e.g., (Fudenberg & Tirole, 1991; Gale & Stewart, 1953; Nash, 1950; Nourani, 1999) is the study of rational (Nash, 1950; Thompson, 2004) behavior in situations in which choices have a mutual effect to one's business and the competitors. Example games businesses play range from price wars and external funding to attrition. Game theory allows you to formulate an effective strategy and can help predict the outcome of strategic situations. Games are often represented with a payoff matrix AND/OR decision trees.

The following section presents new agent game trees the author had put forth (Nourani, 1999). Applying game theory to business is tantamount to an interactive decision theory. Decisions are based on the world as given. However, the best decision depends on what others do, and what others do may depend on what they think you do. Hence games and decisions are intertwined.

10.4.2 SPLITTING AGENT GAME TREES AND BUSINESS PLANNING

A second stage business plan needs to specify how to assign resources with respect to the decisions, ERP plans, and apply that to elect supply chain policies, which can in part specify how the business is to operate. A tactical planning model that plans critical resources up to sales and delivery is a business planner's dream. Planning and tasking require the definition of their respective policies and processes; and the analyses of supply chain parameters. The above are the key elements of a game, anticipating behavior and acquiring an advantage. The players on the business planned must know their options, the incentives, and how do the competitors think.

Example premises: Strategic interactions

Strategies: {Advertise, Do Not Advertise} Payoffs: Companies' Profits

Advertising costs $ million

Advertising captures $ million from a competitor

How can we represent this game?

AND/OR trees (Nilsson, 1969) (Nilsson, 1969; Nilsson, 1971) are game trees defined to solve a game from a player's standpoint. Formally a node problem is said to be solved if one of the following conditions holds.

1. The node is the set of terminal nodes (primitive problem- the node has no successor).
2. The node has AND nodes as successors and the successors are solved.
3. The node has OR nodes as successors and any one of the successors are solved.

A solution to the original problem is given by the subgraph of AND/OR graph sufficient to show that the node is solved. A program which can play a theoretically perfect game would have a task like searching and AND/OR tree for a solution to a one-person problem to a two-person game. Agent game trees were introduced by the author (Nourani, 1994; Nourani, 1996; Nourani & Maani, 2003) with Agent AND/OR trees – an AND/OR tree where the tree branches are based on agent splitting tree decisions (Nourani, 1994; Nourani, 2002). The branches compute a Boolean function via agents. The Boolean function is what might satisfy a goal formula on the tree. An intelligent AND/OR tree is solved if the corresponding Boolean functions solve the AND/OR trees named by intelligent functions on the trees. Thus, node m might be f(a1,a2,a3) & g(b1,b2), where *f* and *g* are Boolean functions of three and two variables, respectively, and ai's and bi's are Boolean-valued agents satisfying goal formulas for *f* and g. The agent trees are applied to satisfy goals to attain competitive models for business plans and ERP (Figure 10.3).

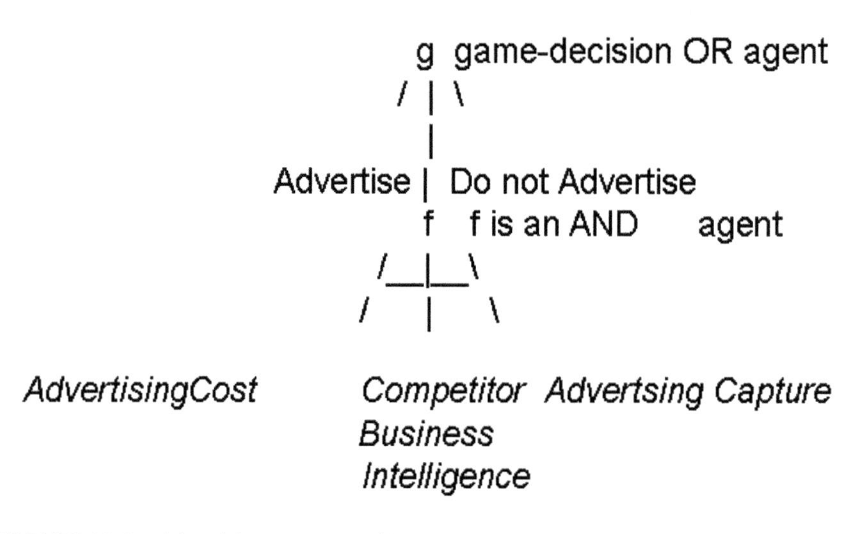

FIGURE 10.3 Advertising agent AND/OR tree.

The AND vs. OR principle is carried out on the above trees with the System to design ERP systems and manage as Cause principle decisions. The agent business modeling techniques the author had introduced (Nourani, 1999; Nourani, 1998) apply the exact 'system as a cause' and 'system as symptom' based on models (assumptions, values, etc.) and the 'system vs. symptom' principle via tracking systems behavior with cooperating computational agent trees. The design might apply agents splitting trees, where splitting trees is a well-known decision tree technique. Surrogate agents are applied to splitting trees. The technique is based on the author's intelligent tree project (Nourani, 1996). The ordinary splitting tree decisions are called CART developed at Berkeley and Stanford by Brieman, Friedman, Olshen, and Stone. The splitting agent decision trees have been developed independently by the author since 1994 at the Intelligent Tree Computing project. For new directions in forecasting and business planning see Nourani (1999); Nourani (2002); and Nourani & Moudi (2005). In Nourani (1999), the author proves upper bounds on infinitary multiplayer games. For example, for every pair p of competing agents, there is a set A<p> where the worse case bound for the number of moves for a determined game based on the agent game tree model is the sum ({|A<p>|: p agent pairs}. At the intelligent game trees, the winning agents determine the specific model where the plan goals are satisfied.

10.4.3 OPTIMAL GAMES?

The following team coding example might be applied as an example to plan optimal games where a project is managed with five tasks/color code. Only the players with the specific matching striped code can be assigned to a task. The goal is to accomplish the five tasks where the goal task is the rectangle with only two balls (Figure 10.4).

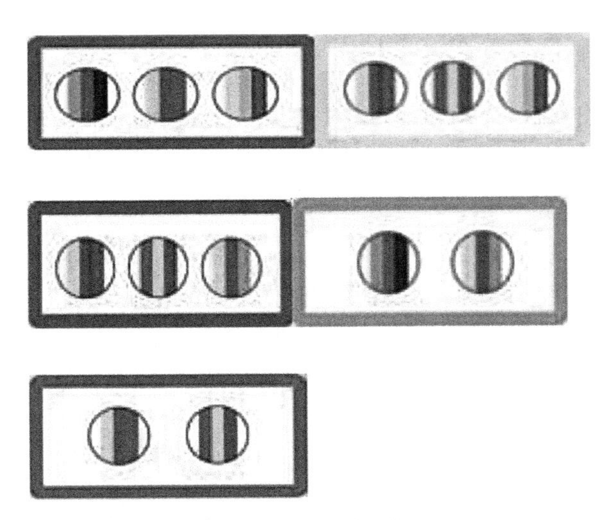

FIGURE 10.4 (See color insert.) Task code optimality.

The competitive optimality principle is to achieve the goals minimizing costs with the specific player code rule. What the techniques presented here are certain bases to address how to acquire an advantage in games on strategic moves, credibility, threats, promises, leveraging limited rationality, and exploiting incomplete information. What is optimality with respect to games? Here are some areas addressed in games: How to anticipate others' behavior in a game. *Dominance:* If never play a strategy that is always worse than another. *Rationalizability:* If play optimal given some beliefs about what others play (and what others believe). *Equilibrium:* If play optimal given correct beliefs about others. Competitive model learning is based on the new agent computing theories we have defined since 1994. Intelligent AND/OR trees and means-end analysis is applied with agents as the hidden-step computations. A novel multiplayer game model is presented where "intelligent" agent enriched languages can be applied to address game questions on models in the mathematical logic sense.

The author has applied game tree planning as a basis to modeling prediction and intent. Applications range from designing complex control systems, to business modeling, economic games, and enterprise resource planning (Nourani, 2002). For intent inference, decision support for teams facing intelligent opponents is limited in their utility without planning systems which can encompass adversary goals and actions. Thus, modeling asymmetric multiplayer competitive games is a key capability to intent inference. The models presented are robust and encompass complex multiple operators. Collaborative agents which have opened new avenues in modeling and implementing teams of cooperative agents are ingredients to specific application modeling, for example, ERP presented in the author's projects. The development of intent-aware decision support for multi-operator complex systems is facilitated with the competitive model's multivalent decision tree computing techniques in the author's projects since 1994. Business planning optimization is a non-linear programming problem, while specific task goals on an ERP assignment are dynamic programming problems with integer programming subproblems. Maximize profits given specific investments and a macro ERP statement without specific tasking costs and pre-specified/task optima. A non-linear program is permitted to have non-linear constraints or objectives.

Portfolio Selection Example: When trying to design financial portfolio investors seek to simultaneously minimize risk and maximize return. The risk is often measured as the variance of the total return, a nonlinear function. The two methods are commonly used: (1) Min Risk s.t. Expected Return

= Bound; (2) Max Expected Return – (Risk), where reflects the tradeoff between return and risk. Regression and estimating Return on Stock A vs. Market Return.

10.5 PLANS AND UNCERTAINTY

Modeling with agent planning is applied where uncertainty is relegated to agents, where competitive learning on game trees determines a confidence interval. The incomplete knowledge modeling is treated with Knowledge representation on predictive model diagrams. Model discovery at Knowledge bases is with specific techniques defined for trees. Model diagrams allow us to model-theoretically characterize incomplete KR (knowledge representation). To key into the incomplete knowledge base, we apply generalized predictive diagrams whereby specified diagram functions a search engine can select onto localized data fields. The predictive model diagrams (Nourani, 1995) could be minimally represented by the set of functions {f1,...,fn} that inductively define the model. Data discovery from KR on diagrams might be viewed as satisfying a goal by getting at relevant data which instantiates a goal. The goal formulas state what relevant data is sought, c.f. Chapter 4. Knowledge representation (KR) has two significant roles: to define a model for the world, and to provide a basis for reasoning techniques to get at implicit knowledge. An ordinary diagram is the set of atomic and negated atomic sentences that are true in a model.

Generic diagrams are diagrams definable by a minimal set of functions such that everything else in the model's closure can be inferred by a minimal set of terms defining the model. Thus, providing a minimal characterization of models, and a minimal set of atomic sentences on which all other atomic sentences depend. A formalization of problem-solving knowledge (we refer to it from now on with the term: theory) is not independently true. Neither is a consequence inferred from a theory true independently. The sets of sentences defining a theory are only true with respect to a certain world. That world is called a model.

10.5.1 PREDICTION AND DISCOVERY

Minimal prediction is an artificial intelligence technique defined in the author's model-theoretic 'planning and deduction with free proof trees' project around 1992–93 at TU Berlin. It is a cumulative nonmonotonic approximation attained with completing model diagrams on what might be true in a model or knowledge base. A *predictive diagram* for a theory T is a diagram D (M), where M is a model for T, and for any formula q in M, either the function f: q o {0,1} is defined, or there exists a formula p in D(M), such that T U {p} proves q; or that T proves q by minimal prediction. Prediction involves constructing hypotheses, where each hypothesis is a set of atomic literals; such that when some particular theory T is augmented with the hypothesis, it entails the set of goal literals G. The hypotheses must be a subset of a set of ground atomic predictable. The logical theory augmented with the hypothesis must be proved consistent with the model diagram. Prediction is minimal when the hypothesis sets are the minimal such sets. A *generalized predictive diagram* is a predictive diagram with D (M) defined from a minimal set of functions. The predictive diagram could be minimally represented by a set of functions {f1,...,fn} that inductively define the model. The free trees we had defined by the notion of probability implied by the definition, could consist of some extra Skolem functions {g1...,gl} that appear at free trees. The *f* terms and *g* terms, tree congruences, and predictive diagrams then characterize partial deduction with free trees. Prediction involves constructing hypotheses, where each hypothesis is a set of atomic literals gi; such that when some particular theory T is augmented with gi, it entails the set of goal literals G, i.e., T U gi logically implies G, written gi $\models G$ (Schöenfeld, 1967) (Van Heijst et al., 1996). gi must be a subset of a set of ground atomic predictable A. In addition, we must ensure T U gi is consistent. The set of all possible hypotheses is g = {gi}. Prediction is minimal when the gi is the minimal such sets. The hypotheses are derived from the discovery data carried on the specific domain.

For specific model diagram examples please refer to Chapters 4 and 7. From the preceding sections example is as follows:

- Intelligent forecasting
- IS-A stock forecasting technique
- Portfolios stock, bonds, and corporate assets
- Member management science techniques

We consider models as Worlds at which the alleged theorems and truths are valid for the world. Models uphold to a deductive closure of the axioms modeled and some rules of inference, depending on the theory. By the definition of a diagram, they are a set of atomic and negated atomic sentences, and can thus be considered as a basis for a model, provided we can by algebraic extension, define the truth value of arbitrary formulas instantiated with arbitrary terms.

A theory for the world reasoned and model is based on axioms and deduction rules, for example, a first-order logical theory. The above scheme with the basic logical rules, can form a basic theory T to reason about a stock forecasting technique. To do predictive analysis we add hypotheses, for example, the atomic literals based on the following:

p1. Asset (Stocks) p2. Stock (x) → Asset (x) p3. S&P500 (x) → Stock (x)

From the definitions a predictive diagram for a theory T is a diagram D[M], where M is a model for T, and for any formula q in M, either the function f: q o {0,1} is defined, or there exists a formula p in D[M], such that T U {p} proves q; or that T proves q by minimal prediction.

The predictive diagram for T is constructed starting with p1 =True, p2(f) = true for *f* ranging over stock symbols, p3 is true for all x=f, where *f* is a stock symbol in S&P500.

To predict if the stock *f* is due to increase in value might be inferred with a predictive diagram that includes hypothesis p3 on the stock symbol and that the average value of S&P500 is due to increase based on the stocks on the specific sector. For example, a predicate p4 that stock symbol f1 is in Sector S1 and the sector S1 is due to an increase.

New hypothesis: let |= be the logical implication symbol

p4. stock (f1) |=S1(f)

p5. S1(x) |= 25 % increase average value (x)

Thus, T U p5 |= increase value (f), i.e., the stock symbol *f* is predicted to appreciate.

An example of the forecasting theory T based on the above is as follows. Defaults, possible hypotheses, that we accept as part of a forecast and observations M which are to be reconciled.

An observation $g \in \varphi$ is reconcilable if there exit ground hypotheses $\gamma \subseteq \Omega$ such that:

1. $T \cup \gamma \models g$; and
2. $T \cup \gamma$ Is consistent.

The author's publications (Nourani & Hoppe, 1994; Nourani, 2003) have proved that a set of first-order observations M is reconcilable with the model iff there exists a predictive diagram for the logical consequences to M.

10.5.2 DATA DISCOVERY AND DATA MODELS

Business intelligence interfaces (Nourani, 2005; Nourani, 2004) might apply automated learning and discovery—often called data mining (Nourani, 2000; Nourani, 2005), machine learning, or advanced data analysis has new interface relevance. There are obvious financial and organizational memory (Nourani, 2005); (Shoenfield, 1967) applications applied at times in our projects. E-commerce, e-business, trust, trustworthiness, usability, Human-Computer Interaction, cognitive ergonomics, user interface design, ease of use, interaction design, and online marketing, are the business user modeling issues the chapter addresses. Financial companies have begun to analyze their customers' behavior in order to maximize the effectiveness of marketing efforts. There are routine applications to data discovery techniques with intelligence databases. Management process controls at times call on warehoused data and rely on organizational memory to reach a decision. Recent research has led to progress, both in the type methods that are available, and in the understanding of their characteristics. The broad topic of automated learning (Nourani, 2005; Sieg, 2001) and discovery are inherently cross-disciplinary in nature. As there is increased reliance on visual data and active visual databases on presenting and storing organizational structures, via the internet and the WWW, the role of data discovery and intelligent multimedia active databases become essential. Knowledge management- KM and organizational memory OM are areas where model discovery with active Intelligent Databases applying predictive logic (Nourani, 1995; Nourani, 2003).

Learning causal relationships amongst stored data is another important area, which can benefit from our project on Intelligent Multimedia business interfaces (Nourani, 2006). Most existing learning algorithms detect only correlations, but are unable to model causality and hence fail to predict the effect of external controls. Interactive discovery data mining is a process which involves automated data analysis and control decisions by an expert of the domain. For example, patterns in many large-scale business system databases might be discovered interactively, by a human expert looking at the data, as it is done with medical data. Data visualization is specifically difficult when data is high-dimensional, specifically when it involves non-numerical data such as text. The projects might be a basis for designing interactive business tools.

10.5.3 THE PLAN OPTIMALITY PRINCIPLE

The goal is satisfied with as a dynamic programming problem where at each decision point a local optimum model is chosen based on non-linear cost criteria. Let us stage a dynamic macro-programming model where the micro stages are local optimization problems on their own right as follows: the LocalMin-Dynamic Model and Computing Algorithm.

Let the start state be S0 and SF be the goal state.

A model is feasible if it satisfies the basic business goals at a stage. For example, meets cost budgets, and assigned resources to accomplish tasks at that stage.

Models M1,..., Mn at stage Sk are competitive feasible models with respect to which and a non-linear cost criterion a local optimum is computed with the following algorithm.

State S0

Initial state: Capital, budgets, and expenses, projected profits based on a static business plan.

Start from the projected goal state SF.

Apply the non-linear cost criterion which is based on computing business model where with respect to computing confidence intervals to select a model that satisfies the local business goals based on a time period, e.g., a quarter performance, to take back state SF–1, Iterate to reach an optimum state S0.

Minimal prediction is applied from the preceding sections to select minimum cost models based on uncertainty. There may be several local optimal models. Let x be a feasible solution, then –x is a global max if f(x) = f(y) for every feasible y. –x is a local max if f(x) = f(y) for every feasibly sufficiently close to x (i.e., xj–e = yj = xj + e for all j and some small e).

10.5.4 EXAMPLE OPTIMAL MODEL

Minimize Non-linear cost criteria cost = capital – earnings – (quarter expenses) Projected goal state SF Annual gross 1/2 M.

Must select models that minimize cost criteria and have at least ½ M annual gross.

Model	Capital	Staff numbers	Annual expenses	Quarter earnings	Projected annual gross
50 K/staff					
M1	1 M	10	500 K	250 K	1 M
M2	1 M	7	350 K	300 K	1.2 M
M3	1 M	8	400 K	200 K	0.8 M
M4	1 M	6	300 K	190 K	0.75 K

Compute K for M1–M4:

Select model for SF–1

M1: cost = 1M – 250,000–125,000 = 625,000
M2: cost =1M – 300,000–87,500 = 612,500
M3: cost = 1M – 200,000–100,000 = 700,000
M4: cost = 1M – 190,000–75,000 = 735,0000

State SF–1 takes model M2 so that the new capital is 612,500

M1: cost = 612,500–250,000–125,000 = 237,500
M2: cost = 612,500–300,000–87,500 = 225,000
M3: cost = 612,500–200,000–100,000 = 312,500
M4: cost = 612,500–1,900,000–75,000 = 347,500

State SF–2 takes model M3 with a new capital 312,500

M1: cost = 312,500–250,000–125,000 = negative
M2: cost = 312,500–300,000–87,500 = negative
M3: cost = 312,500–200,000–100,000 = 12,500
M4: cost = 312,500–190,000–750,000 = negative

State SF–3 takes M3, where the feasibility limit is reached for the alternate models. The optimal models that are used at each stage are M2, M3, M3. The optimal model to use is M3.

10.6 APPLICATIONS TO E-COMMERCE – MANAGEMENT FLIGHT SIMULATORS

We can examine a specific management flight simulator to exhibit an EC example to the above competitive modeling techniques. Example management simulations are *Project Challenge,* a simulation designed to provide the player with the opportunity to exercise management judgment in a realistic information technology project environment. The focus is on managing a project; the player experiences the complex inter-dependencies and tradeoffs of management decisions that affect the project's schedule, budget, team morale, and customer satisfaction. Players can interview team members, listen in on gossip in the lunchroom, and go ask the boss for more money. *The Information Technology Organization Flight Simulator* Margaret Johnson's experiment at Stanford Computer Science groups (Johnson, 1992–1993) since mimics the operation of project management in a small MIS department. Players make decisions in areas such as accepting new project work, hiring additional personnel, and investment in technology infrastructure. Developing EC systems for (Johnson, 1992–1993; Senge, 1992) enterprises have to plan their resources carefully, particularly in their "incubation" phases to ensure profitability for their members. There are typically three types of resources to manage:

1. Business development – responsible for identifying and developing sales opportunities.
2. Member recruitment – responsible for recruiting and inducing (the right kind of) new members and companies.
3. Network development – responsible for building shared ambitions, values, network governance, and business infrastructures.

The Microworld management flight simulator illustrates competitive business models. The above simulators have real limits on their predictive ability. All specific models can be wrong on the particulars. We can never be certain apriori that the models mimic reality. For example, a driver's reaction to an icy patch might be counterintuitive, and the typical driver might do the exact wrong thing to turning into the skid. The only value of a model is its utility or usefulness. That is why our predictive competitive modeling is a route to choose to obtain the most accurate approximation to the real outcome. At each stage, the model that optimizes the criteria is selected by the algorithm according to the preceding sections. The five basic rules of the Virtual EC systems are:

1. "Opportunities to Bid";
2. "Contracts Won" at a rate which depends on the number of members and the social capital/relationships, collaboration skills and

infrastructure capital developed. That is, the ability to close contracts depends on the perceived capability of the system to deliver.

With our competitive modeling technique based on certain initial conditions, for example, 0 contracts won. There are four functions corresponding to the startup variables based on the discovered data, for example, we can explore n models based on altering the EC Network Capital assignments, and an opportunity to bid and breadth-depth parameters. Optimizing the models a promising model emerges that meets the specifications. Since the author's 1994 papers on model diagrams, agent games, and decisions, there are recent applications to games on influence diagrams and Bayesian causal trees at Stanford (Koller & Milch, 2001), the relationship to the author's 1994–2007 projects might yet be explored.

10.7 CONCLUSIONS

A Generalized MinMax local optimum dynamic programming algorithm was presented and applied to business model computing where predictive techniques can determine local optima. Specific business planning and EC area applications were introduced. Planning, goals, prediction schemes and open loop control principles are stated in a new dynamic programming algorithm. ERP is modeled and realized with competitive models, learning agents, and goals. Splitting agent game trees are introduced and applied. Optimal games are contemplated and examined. The competitive optimality principle is to achieve the goals minimizing costs with the specific player code rule. What the techniques presented here accomplish are certain bases to techniques that can allow us to acquire an advantage in games on strategic moves (Dixit & Skeath, 1999). The benefits to the business practice are to have practical optimal business plans, for example, ERP specifics, and practical predictive modeling techniques, to name a few. Model discovery and prediction are applied to compare models and get specific confidence intervals to supply to goal formulas. Competitive model learning is based on the new agent computing theories the author defined since ECAI 1994. Newer applications for the areas developed are innovations management. The computing model is based on a novel competitive learning with agent multiplayer game tree planning. Specific predictive applications and optimization models were presented. Modeling with agent planning is applied where uncertainty is relegated to agents, where competitive learning on game trees determines a move. The incomplete knowledge modeling is treated on predictive model diagrams but the exact realization takes considerable development with domain specific problem areas to address. There are new practical and theoretical areas they can benefit from further research. The competitive modeling area has to be further developed to tune the specific variables and parameters. Game applications to optimal plans are further areas that deserve considerable research. A full treatment is beyond the scope of the specific exposition here.

KEYWORDS

- **agent game trees**
- **competitive model dynamic programming**
- **competitive models**
- **multiplayer games**
- **planning**
- **predictive modeling**

REFERENCES

Anderson, V., & Johnson, L., (1997). *Systems Thinking Basics*. Pegasus Communications, Inc.

Bellman, R. E., (1975). *Dynamic Programming*. Princeton University Press, Princeton, NJ, USA.

Bertsimas, D., & Freund, R., (2000). *Data, Models, and Decisions: The Fundamentals of Management Science*. Southwestern College Publishing.

Buffett, W., (2000). *The Billionaire's Buyout Plan*. The New York Times.

Bullard, J., & Duffy, J., (1998). A model of learning and emulation with artificial adaptive agents, *Journal of Economic Dynamics and Control,* Elsevier, *22*(2), pp. 179–207.

Chevalier, J., (1999). The Pros and Cons of Entering a Market. *Financial Times*, 8–10.

Dixit, A. K., & Skeath, S., (1999). *Games of Strategy*. New York, NY: W. W. Norton & Company. ISBN: 0393974219.

Fleisher, C. S., & Bensoussan, B., (2007). *Business and Competitive Analysis: Effective Application of New and Classic Methods* (1st edn., p. 365–371, 528). ISBN-10: 0-13-187366-0; ISBN-13: 978-0-13-187366-7; Published: Feb 27, 2007, Copyright; Dimensions 7x9-1/4.

Forrester, J., (1958). *Industrial Dynamics – A major Breakthrough for Decision Makers*. Harvard Business Review, *36*(4).

Fudenberg, D., & Tirole, J., (1991). *Game Theory*. MIT Press.

Gale, D., & Stewart, F. M., (1953). Infinite games with perfect information, in contributions to the theory of games, *Annals of Mathematical Studies, 28*, Princeton.
Hagel, J. V., & Armstrong, A. G., (1997). *Net Gain-Expanding Market Through Virtual Communities*. Harvard Business School Press, Boston, Mass.
Holland, J. H., & Miller, J. H., (1991). Artificial adaptive agents in economic theory, American economic association. *Journal American Economic Review, 81.*
Johnson, (1992–1993). *Margaret Systems Thinking and the I/T Professional, a Monthly Column in Information Technology and Measurement*. Cutter Publications.
Kauffman, D. L. Jr., (1980). Systems one, an introduction to systems thinking, The Innovative Learning Series, Future Systems, Inc.
Koller, D., & Milch, B., (2001). Structured models for multi-agent interactions, theoretical aspects of rationality and knowledge. *Archive Proceedings of the 8th Conference on Theoretical Aspects of Rationality and Knowledge Siena* (pp. 233–248). Italy. Session: Computer science, games, and logic, Year of Publication: 2001, ISBN: 1-55860-791-9.
Maani, K., & Cavana, R., (2000). *Systems Thinking and Modeling – Understanding Change and Complexity*, Prentice Hall.
Miller, J. H., Butts, C. T., & Rode, D., (2002). Communication and cooperation. *Journal of Economic Behavior & Organization,* Elsevier, *47*(2), 179–195.
Nash, J., (1950). Equilibrium points in n-person games. *Proc. National Academy of Sciences of the USA, 36*, 48–49.
Nilsson, D., & Lauritzen, S. L. (2000). Evaluating influence diagrams using LIMIDs. *Proceedings of the 16th Conference on Uncertainty in Artificial Intelligence*, pp. 436–445.
Nilsson, N. J., (1969). Searching, problem-solving, and game-playing trees for minimal cost solutions. In Morell, A. J., (ed.), *IFIP 1968* (Vol. 2, pp. 1556–1562), Amsterdam, North-Holland.
Nilsson, N. J., (1971). *Problem Solving Methods in Artificial Intelligence*, New York, McGraw-Hill.
Nourani, C. F. (1997). *AII Applications to Heterogenous Design, International Conference on Multiagent AI*, University of Karlskrona/Ronneby, Department of Computer Science and Business Administration Ronneby, Sweden, May. www recording only.
Nourani, C. F., & Hoppe, T., (1994). *GF-Diagrams for Models and Free Proof Trees, Technical Universitat Berlin, Informatics*, Berlin Logic Colloquium, Universitat Potsdamm, Potsdamm, Germany. Humboldt Universtiat Mathematik sponsor.
Nourani, C. F., & Maani, K., (2003). *Enterprise Planning and Systems Modeling Two Synergistic Partners*. SSGRR, Rome, December. The University of Auckland MSIS, Invited Paper.
Nourani, C. F., & Moudi, R. M., (2005). *Open Loop Control and Business Planning: A Preliminary Brief.* International Conference Autonomous Systems, Tahiti, French Polynesia. October 23–28, 2005. www.iaria.org/conferences/ICAS/ICAS2005/ GeneralInformation/ GeneralInformation.html, page 8- ISBN: 0–7695–2450–8.
Nourani, C. F., (1994). *A Theory for Programming With Intelligent Syntax Trees And Intelligent Decisive Agents- Preliminary Report 11th European Conference A.I.*, ECAI Workshop on DAI Applications to Decision Theory, Amsterdam.
Nourani, C. F., (1995). *Free Proof Trees and Model-theoretic Planning.* Automated Reasoning, AISB, England.
Nourani, C. F., (1996). *Slalom Tree Computing, 1994.* AI Communications, IOS Press, Amsterdam, ISSN 0921-7126.

Nourani, C. F., (1997). *Intelligent Tree Computing, Decision Trees and Soft OOP Tree Computing*, Frontiers in soft computing and decision systems, Papers from the 1997 fall symposium, Boston, Technical Report FS-97-04 AAAI, ISBN: 1-57735-079-0, www.aaai.org/Press/Reports/Symposia/Fall/fs-97-04.html.
Nourani, C. F., (1998). *Agent Computing and Intelligent Multimedia – Challenges and Opportunities for Business and Commerce*. Preliminary. Project M2 & the University of Auckland Management Science.
Nourani, C. F., (1998). Agent computing, KB for intelligent forecasting, and model discovery for knowledge management. *AAAI Workshop on Agent-Based Systems in the Business Context*. Orlando, Florida, AAAI Press, Menlo Park, CA.
Nourani, C. F., (1999). *Business Modeling and Forecasting*. AIEC-AAAI99, Orlando, AAAI Press.
Nourani, C. F., (1999). *Infinitary Multiplayer Games, Summer Logic Colloquium*. Utrecht, BSL, ASL Publications.
Nourani, C. F., (1999). *Management Process Models and Game Trees Applications to Economic Games*. Invited Paper SSGRR, L'Auquila, Rome, Italy.
Nourani, C. F., (2002). *Game Trees, Competitive Models, and ERP*. New Business Models and Enabling Technologies, Management School, St Petersburg, Russia, Keynote Address. Fraunhofer Institute for Open Communication Systems, Germany 11:00–12:00 www.math.spbu.ru/user/krivulin/Work2002/Workshop.htm. Proceedings Editor Nikolai Krivulin.
Nourani, C. F., (2002). *Relevant World Models, and Model Discovery, (1997)*. Brief published at AKA99, Australia AI, Sydney.
Nourani, C. F., (2003). *Predictive Model Discovery, and Schema Completion*. World Multiconference on Systemics, Cybernetics, and Informatics (SCI 2002) Orlando, USA, http://www.iiis.org/sci2002.
Nourani, C. F., (2004). *W-Interfaces, and Business Intelligence with Multitier Designs CollECTeR LatAm*. Santiago, Chile, http://ing.utalca.cl/collecter/techsession.php.
Nourani, C. F., (2005). *Intelligent Multimedia Computing Science: Business Interfaces, Wireless Computing* (p. 200). Databases, and Data Mining, CA, Hardcover, ISBN: 1–58883–037–3.
Nourani, C. F., Grace, S. L., & Loo, K. R., (2000). *Model discovery from active DB with Predictive logic, Data Mining 2000 Applications to Business and Finance Cambridge*, UK.
Nourani, C. F., (2016). *Ecosystems and Technology Idea Generation and Content Model Processing* (pp. 282 with index). Apple Academic Press. ISBN hard: 978-1-77188-507-2. Cat #: N11817 ISBN eBook: 978–1–315–36566–4. Cat #: NE11982.
Nourani, C.F., & Lauth, L., (2015). Predictive control, competitive model business planning, and innovation ERP. ICCMIT, Prague, *Procedia Computer Science*, pp. 891–900. Elsevier Publishers.
Parsons, et al. (2002). In: Parsons, S., Gymtrasiewicz, P., & Wooldridge, M., (eds.), *Game Theory and Decision Theory in Agent-Based Systems Series: Multiagent Systems, Artificial Societies, and Simulated Organizations* (Vol. 5, p. 416). Hardcover ISBN–10: 1–4020–7115–9, ISBN–13: 978–1–4020–7115–7, Springer-Verlag.
Peterson, I., (1996). *Mating Games and Lizards*. Mathematical Association of America.
Riggins, F. J., & Sue Rhree, H. S., (1998). Towards a unified view of E-commerce. *CACM*, *41*(10), 88–95.
Rubin, H. M., Johnson, & Yourdon, E., (1994). *Process Flight Simulation, Managing System Development*, Cutter Publications.

Senge, P., (1992). Building learning organizations, *Journal for Quality and Participation.*
Shoenfield, J. R. (1967). *Mathematical Logic*, Addison Wesley, ISBN 0-201-07028-6.
Sieg, G., (2001). A political business cycle with boundedly rational agents. *European Journal of Political Economy*, Elsevier, *17*(1), 39–52.
Thompson, K., (2004). *A Management Flight Simulator for Virtual Enterprise Network Incubation.* 10th International Conference on Concurrent Enterprising, ICE2004 Sevilla, Spain.
Van Heijst, G., Van der Speck, R., & Kruizinga, E., (2000). Organizing corporate memories. In: *Proceedings of KAW*. http://ksi.cpsc.ucalgary.ca/KAW/KAW96/KAW96Proc.html.03.

CHAPTER 11

COMPETITIVE MODELS, COMPATIBILITY DEGREES, AND RANDOM SETS

CYRUS F. NOURANI

TU Berlin AI, Berlin, Germany, and SFU, Burnaby, BC, Canada

ABSTRACT

Chapter examines random sets as a basis to carry structures modeling towards a competitive culmination problem where models "compete" based on modeling game trees. A model rank is higher when on game trees with a higher game tree degree, satisfies goals, hence realizing specific models where the plan goals are satisfied. Characterizing competitive model degrees on random sets is a basic area to explore. A model is a competing model iff at each stage the model is compatible with the goal tree satisfiability criteria. Compatibility is defined on Random Sets where the correspondence between compatibility on random sets and game tree degrees are applied to present random model diagrams. Random diagram game degrees are applied and model ranks based on satisfiability computability to optimal ranks are examined.

11.1 INTRODUCTION

Games play is an important role as a basis to economic theories. Here, the import is brought forth onto decision tree planning with agents. The author had presented specific agent decision tree computing theories since 1994 and can be applied to present precise strategies and prove theorems on multi-player games. Game tree degree with respect to models is defined and applied to prove soundness and completeness. The game is viewed as a multiplayer game with only perfect information between agent pairs. Upper bounds on determined games were presented on first authors publications stated. A technique for modeling game tress satisfiability is based on competitive models (Nourani, 2008) and Section 11.2. The present chapter is a preliminary basis to carry on competitive model satisfiability as a basis to optimize decisions based on random sets (Martin-Löf, 1966; Nourani, 2009).

Random sets are random elements taking values, as subsets of some space, are mathematical models for set-valued observations and irregular geometrical patterns. Random sets in stochastic geometry (Kendall, 1974) are examples. Besides sampling designs, confidence regions, stochastic geometry, and morphological problems, random sets appear in general as set-valued observed processes. The concept of random sets has been carried onto random fuzzy sets to model perception-based information in social systems, artificial intelligence problems such as intelligent control and decisions. Our specific chapter addresses the question: how to model competitive model computing planning with random sets where a novel model is developed with diagrammatic techniques to carry out competitive model planning to reach specific goals or to carry on game tree objectives. Section begins to develop competitive model game trees brief from the first author's recent publications. Section 11.3 starts to characterize competitive model degrees based on random sets, where on section form non-deterministic diagrams are applied to compatibility on models. Computations geometry on random algorithms is previewed to projections on Boolean-valued maps to product random sets. Model ranks are presented based on random model diagrams. Section 11.5 defined random diagram game trees where computability questions on model compatibly are addressed and model ranking complexity is examined. The authors have applied the competitive modeling techniques to game tree modeling on few publications since 2012 (Nourani, Schulte, 2012, 2013).

In the author's earlier chapters that of the definition of generalized nondeterministic diagrams captures the method of Possible Worlds. Further, the earlier notion of a set {T,F,X} in Tourane (1991) and diagrams with generalized Skolemization (Nourani, 1993a, 1995b) handle arbitrary valued logic. If we were to search for a model-theoretic (Nourani, 1991) view of these in view of a triple <L,|A|,A> for a proper language L, we could gain some insight to the approaches. The other components of the triple are the concept of a model A and its universe |A|, see for example (Kleene, 1952). The diagram of a structure is the set of atomic and negated atomic sentences that are true in that structure.

The generalized diagram (G-diagram) (Nourani, 1987, 1991) is a diagram in which the elements of the structure are all represented by a minimal family of function symbols and constants, such that it is sufficient to define the truth of formulae only for the terms generated by the minimal family of functions and constant symbols. c.f. Chapters 4 and 7. This allows us to define a canonical model of a theory in terms of a minimal family of function symbols In a possible world approach one focuses on the "state of affairs" that are compatible with what one knows to be true.

11.2 COMPETITIVE MODELS AND GAME TREES

Planning is based on goal satisfaction at models. We can examine random sets as a basis to carry structures modeling as a competitive culmination problem where models "compete" based on modeling game trees, where the model rank is higher when an on game trees with a higher game tree degree, satisfies goals, hence realizing specific models where the plan goals are satisfied.

When a specific player group "wins" to satisfy a goal the group has presented a model to the specific goal, presumably consistent with an intended world model. c.f. Chapters 4 and 10. The intelligent trees (Nourani, 1994, 1996) are ways to encode plans with agents and compare models on goal satisfaction to examine and predict via model diagrams why one plan is better than another or how it could fail. Virtual model planning is treated in the author's publications where plan comparison can be carried out at VR planning (Nourani, 1999b).

11.2.1 MODELS PRELIMINARIES

To have precise statements on game models we begin with basic model-theoretic preliminaries. Agent models and morphisms on agent models had been treated in the first author's publications since the past decade. We develop that area further to reach specific game models. We apply basic model \theory techniques to check upward compatibility on game tree goal satisfaction.

The modeling techniques are basic game tree node stratification. Since our game tress models have "categorical axiomatization" we can further explore game model embedding techniques with morphisms that the author has developed on categorical models (Nourani1996, 2005). Let us start from certain model-theoretic premises with propositions known form basic model theory. A *structure* consists of a set along with a collection of finitary functions and relations, which are defined on it. Universal algebra studies structures that generalize the algebraic structures such as groups, rings, fields, vector spaces, and lattices.

Model theory has a different scope than universal algebra, encompassing more arbitrary theories, including foundational structures such as models of set theory. From the model-theoretic point of view, structures are the objects used to define the semantics of first-order logic. A *structure* can be defined as a triple consisting of a *domain A,* a signature Σ, and an *interpretation function I* that indicates how the signature is to be interpreted on the domain. To indicate that a structure has a particular signature σ one can refer to it as a σ-structure. The domain of a structure is an arbitrary set (non-empty); it is also called the *underlying set* of the structure, its *carrier* (especially in universal algebra), or its *universe* (especially in model theory).

Sometimes the notation or is used for the domain of, but often no notational distinction is made between a structure and its domain, i.e., the same symbol can refer to both. The signature of a structure consists of a set of *function symbols* and *relation symbols* along with a function that ascribes to each symbol *s* a natural number, called the *arity* of *s.* For example, let *G* be a graph consisting of two vertices connected by an edge, and let *H* be the graph consisting of the same vertices but no edges. *H* is a subgraph of *G,* but not an induced substructure. Given two structures and of the same signature Σ`, a *Σ-homomorphism* is a map that can preserve the functions and relations.

11.2.2 MODEL EMBEDDINGS

A Σ-homomorphism h is called a Σ-embedding if it is one-to-one and for every *n-ary* relation symbol *R* of Σ and any elements ai...,an, the following equivalence holds: R(a1...,an) iff R(h(a1).h(an)). Thus, an embedding is the same thing as a strong homomorphism, which is one-to-one. A structure defined for all formulas in the language consisting of the language of A together with a constant symbol for each element of *M,* which is interpreted as that element. A structure is said to be a *model* of a theory *T* if the language of M is the same as the language of *T* and every sentence in *T* is satisfied by M. Thus, for example, a "ring" is a structure for the language of rings that satisfies each of the ring axioms, and a model of ZFC set theory is a structure in the language of set theory that satisfies each of the ZFC axioms.

Two structures M and N of the same signature σ are *elementarily equivalent* if every first-order sentence (formula without free variables) over σ is true in M if and only if it is true in N, i.e., if M and N have the same complete first-order theory. If M and N are elementarily equivalent, written $M \equiv N$. A first-order theory is complete if and only if any two of its models are elementarily equivalent. An *elementary embedding* of a structure N into a structure M of the same signature is a map $h: N \rightarrow M$ such that for every first-order σ- formula φ(x1…,xn) and all elements *a1…,an* of N, $N \models$ φ(a1,..., an) implies $M \models$ φ(h(a1)..., h(an)). Every elementary embedding is a strong homomorphism, and its image is an elementary substructure.

11.3 CHARACTERIZING COMPETITIVE MODEL DEGREES ON RANDOM SETS

Let us start with certain basic premises. Let us say that *a model is a competing model iff at each stage the model is compatible with the goal tree satisfiability criteria.*

In computational geometry, a standard technique to build a structure like a convex hull is to randomly permute the input points and then insert them one by one into the existing structure. The randomization ensures that the expected number of changes to the structure caused by an insertion is small, and so the expected running time of the algorithm can be upper bounded. This technique is called randomized incremental construction. Graph problems are another area that randomized algorithms are applied, for example, a randomized minimum cut algorithm:

find_min_cut(undirected graph G)

{while there are more than 2 nodes in G do {pick an edge (u,v) at random in G contract the edge, while preserving multi-edges remove all loops}

output { the remaining edges }

There are various notions of algorithmic randomness. Relativized randomness is as follows: A set is random if it is Martin-Löf random relative to $(n - 1)$. For example, Nies et al. (2005) show that a set is 2-random if and only if there is a constant c such that infinitely many initial segments x of the set are c-incompressible. *Let us develop model compatibility on random sets and game tree degrees. We begin with model* computing on structures that are definable with certain basic functions and constants. That is an abstract computing that is carried on with functions that are computable, thus structures that are defined by computable functions. When you have such structures you can check what is true on the structure with respect to statements stated with first-order logic, evaluation, or Horn clauses that can take a free assignment to variables to logical formulas that can be stated on tree terms defined with the computable functions. That brings us to two definitions.

Definition 3.1: Let M be a structure for a language L, call a subset X of M a generating set for M if no proper substructure of M contains X, i.e., if M is the closure of X U

{c[M]: c is a constant symbol of L}. An assignment of constants to M is a pair <A,G>,

where A is an infinite set of constant symbols in L and G: A $\rightarrow$ M, such that {G[a]: a in A} is a set of generators for M. Interpreting a by g[a], every element of M is denoted by at least one closed term of L[A]. For a fixed assignment <A,G>of constants to M, the diagram *of M, D<A,G>[M]* is the set of basic [atomic and negated atomic] sentences of L[A] true in M. [Note that L[A] is L enriched with set A of constant symbols.].

Generic diagrams, denoted by G-diagrams, were what the first author had defined over a decade ago 1980's to be diagrams for models defined by a specific function set, for example, $\Sigma 1$ Skolem functions.

Definition 3.2: A generic model diagram is a diagram where D<A,G> on definition 3.1 is characterized with specific functions, e.g., specific Skolem functions.

The following proposition is well-known in model theory.

Proposition 1: Let A and B be models for a language L. Then, A is isomorphically embedded in B iff B can be expanded to a model of the diagram of A.

Let us now define a new compatibility criterion.

Definition 3.3: Two generic diagrams D1 and D2 on the same logical language £ are compatible iff forever assignment to free variables to models on £, every model M for £, M ⊨ D1 iff M ⊨ D2.

The computing specifics are based on creating models from generic model diagram functions where basic models can be piece-meal designed and diagrams completed starting from incomplete descriptions at times. Models uphold to a deductive closure of the axioms modeled and some rules of inference, depending on the theory. By the definition of a diagram, they are a set of atomic and negated.

11.4 RANDOM SETS AND NONDETERMINISM

11.4.1 COMPUTATIONAL MODEL SPECIFICS

By the definition of a diagram, they are a set of atomic and negated atomic sentences. Thus, the diagram might be considered as a basis for a model, provided we can by algebraic extension, define the truth value of arbitrary formulas instantiated with arbitrary terms. Thus, all compound sentences build out of atomic sentences then could be assigned a truth-value, handing over a model.

This will be made clearer in the following subsections. Consider the primitive first order language (FOL) L = {c},{f(X)},{p(X),q(X)}. Let us apply Prolog notation convention for constants and variables) and the simple theory {for all X: p(X) q(X),p(c)}, and indicate what is meant by the various notions. [model] = {p(c),q(c),q(f(c)),q(f(f(c))).},{p(c) &q(c), p(c) & p(X), p(c) &p(f(X))}, {p(c) v p(X), p(c) v p(f(X)), p(c) – p(c).}

[diagram] = {p(c),q(c),p(c),q(f(c)),q(f(f(c))).}.,q(X)}, i.e., diagram = the set of atomic formulas of a model. Thus, the diagram is [diagram] = {p(c), q(c),q(f(c)),q(f(f(c))),q(X)} i.e., diagram = the set of atomic formulas of a model Thus the diagram is [diagram] = {p(c),q(c),q(f(c)),q(f(f(c))),.,q(X)}.

Based on the above, we can define generalized diagrams. The term generalized is applied to indicate that such diagrams are defined by algebraic extension from basic terms and constants of a language. The fully defined diagrams make use of only a minimal set of functions. Generalized diagram is [generalized diagram] = {p(c),q(c),p(f(t)),q(f(t))} for t defined by induction, as {t0 = c, and tn = {f(t(n – 1))} for n > 0. It is thus not necessary to redefine all f(X)'s since they are instantiated. Nondeterministic diagrams are those in which some formulas are assigned an indeterminate symbol, neither true nor false, that can be arbitrarily assigned in due time. [nondeterministic diagram] = {p(c),q(c),p(f(t)),q(f(c)),q(f(f(c))),I_q(f(s)))}, t is as defined by induction before and I_q(f(s)) = I_q for some indeterminate symbol I_q, for

{s = tn, n >= 2}.

These generic diagrams are applicable for knowledge representation in planning with incomplete knowledge with backtracking. Free Skolemized diagrams are those in which instead of indeterminate symbols there are free Skolem functions that could be universally quantified. [free Skolemized diagram] = {p(c),q(c),p(f(t)),q(f(c)),q(f(f(c))),q_F.s(s)}, where t and s are as defined in the sections before. A nondeterministic diagram is a diagram with indeterminate symbols instead of truth-values for certain formulas.

For example, [nondeterministic diagram] = {p(c),q(c),p(f(t)),q(f(c)),q(f(f(c))),I_q(f(s)))}, t is as defined by induction before and I_q(f(s)) = I_q for some indeterminate symbol I_q, for {s=t sub n, n>=2}. p and q are language predicate letters and I_q is a variable letter designated by the index letter q to indicate Boolean model assignments, e.g., {T,F} only.

Computing models are structures where certain computation properties stated on a logic on the functions defining the structures can be ascertained true or false. Let us consider computing on structures on a specific signature that can take assignments to variables from a random set. We can rank formulas with ranking based on a measure rank: Formulas with ranking less than a specific value would be assigned 'T' and the other formulas would be assigned 'F.' Corresponding to Possible Worlds we can define Random worlds where we can consider a formula to be true iff it is true with all such random assignments.

This author had defined the notion of a plausible diagram, which can be constructed to define plausible models for revised theories. In practice, one may envision planning with plausible diagrams such that certain propositions are deliberately left indeterminate to allow flexibility in planning. Such extensions to the usual notion of the diagram in model theory are put forth around 1987. That approach was one method of avoiding the computational complexity and computability problems of having complete diagrams. Truth maintenance and model revision can all be done by a simple reassignment to the diagram. The canonical model of the world is defined directly from the diagram. Corresponding to the above we have: Random Model diagrams to compute *compatibility defined on Random Sets.*

A possible world may be thought of as a set of circumstances that might be true in an actual world. The possible worlds analysis of knowledge began with the work of Hintikka through the notion of model set and (Kripke 63) through modal logic, in which rather than considering individual propositions, one focuses on the `state of affairs' that are compatible with what one knows to be true, rather than being regarded as possible, relative to a world believed to be true, rather than being absolute. For example, a world w might be a possible alternative relative to w,' but not to w." Possible worlds consist of a certain completeness property: for any proposition p and world w, either p is true in w or not p is true in w.

Note that this is exactly the information contained in a generalized diagram, as defined in the previous section. Let W be the set of all worlds and p be a proposition. Let [p] be the set of worlds in which p is true. We call [p] the truth-set of P. Propositions with the same truth-set are considered identical. Thus, there is a one-one correspondence between propositions and their truth sets. Boolean operations on propositions correspond to set-theoretic operations on sets of worlds. A proposition is true in a world if and only if the particular world is a member of that proposition.

11.4.2 SITUATIONS AND COMPATIBILITY

From the first author presented a descriptive epistemology starting 1994 ASL, e.g., (Nourani, 2009). Now let us examine the definition of the situation and view it in the present formulation. Due to the above truth-values, definition 3.1, the number of situations exceeds the possible worlds are situations with no missing. The possible worlds being those situations with no missing information and no contradictions. From the above definitions, the mapping of terms and predicate models extend as in standard model theory. Next, a compatible set of situations is a set of situations with the same domain and the same mapping of function letters to functions.

In other words, Definition 3.2 has a proper definition of specific function symbols, e.g., Σ1-Skolem functions. What it takes to have an algebra and model theory of epistemic states, as defined by the generalized diagram of possible worlds is exactly what (Nourani 1998, 91) had accomplished to decide compatibility of two situations we compare there generalized diagrams. Thus, we have the following theorem.

Theorem 4.1: Two situations are compatible iff their corresponding generic diagrams are compatible with respect to the Boolean structure of the set to which formulas are mapped (by the function h above, defining situations).

Proof: The generic diagrams, definitions above encode possible worlds and since we can define a one-to-one correspondence between possible worlds and truth sets for situations, computability is definable by the generic-diagrams.

Towards applying model diagram techniques to geometric computability from the author's descriptive computing (Nourani, 2009) we brief here the following.

Applying generic diagrams to define knowledge with Skolem functions the computation the Skolemized formulas are defined and expanded to assign truth values to the atomic formulas. Boolean algebras can be defined corresponding to the formulas with Skolemized trees. A truth valuation can be defined with respect to a class of Boolean algebras R as a homomorphism from a closure of formulas to an algebra in R.

For example we can define B = <B,min,max,-,0,1> where min and max are defined by induction on depth of Skolemized trees. By the theorem below from Richter (1978), the logical implication for formulas and models for classes of Boolean algebras can be reduced to only the Boolean algebra over {0,1}.

Theorem 4.2: If R is the class of Boolean algebras and R' contains only the Boolean algebra over {0,1} then |= for R is equivalent to |= for R.'

The idea is that if the free proof tree is constructed then the plan has a model in which the goals are what it implies for Skolemized trees and Boolean algebras are the computation by theorem 4.2.

Theorem 4.3: There is a Boolean algebra R' in the class R, R' containing only {0,1}, such that for the standard model A defined by the generic diagram on the compatibility theorem. A |= is equivalent to |= for R.'

Proof: follows from Richter (1978). The following proposition has a glimpse of applications, while an outline, the detail can well comprise a dissertation for prospective authors.

Proposition 2: Computational geometry with random sets can be characterized as a projection with a Boolean-valued function on random diagrams on the corresponding geometry to a product random set.

Proof: (outline): From Section 11.3 example that in computational geometry, a standard technique to build a structure like a convex hull is to randomly permute the input points and then insert them one by one into the existing structure. The structure models the random assignments to a nondeterministic model diagram, that with definitions 3.1 and Theorem 4.1, becomes consistent with the goal structure, e.g., a convex hull. The product characterizations follow from product structure modeling techniques, e.g., Nourani (2005, 2008, 2009).

11.5 RANKS, MODELS, AND RANDOM DIAGRAMS

Based on game trees on competitive models: AND/OR trees (Nilsson, 1969) are game trees defined to solve a game from a player's standpoint. Formally a node problem is said to be solved if one of the following conditions hold.

1. The node is the set of terminal nodes (primitive problem - the node has no successor).
2. The node has AND nodes as successors and the successors are solved.
3. The node has OR a node as successors and any one of the successors is solved.

A solution to the original problem is given by the subgraph of AND/OR graph sufficient to show that the node is solved. A program which can play a theoretically perfect game would have a task like searching and AND/OR tree for a solution to a one-person problem to a two-person game.

An intelligent AND/OR tree (Nourani, 1996) is where the tree branches are intelligent trees. The branches compute a Boolean function via agents. The Boolean function is what might satisfy a goal formula on the tree. An intelligent AND/OR tree is solved iff the corresponding Boolean functions solve the AND/OR trees named by intelligent functions on the trees. Thus, node m might be f(a1,a2,a3) & g(b1,b2), where *f* and *g* are Boolean functions of three and two variables, respectively, and ai's and bi's are Boolean-valued agents satisfying goal formulas for *f* and *g* (Figure 11.1).

The chess game trees can be defined by agent augmenting AND/OR trees (Nilsson, 1969). For the intelligent game trees and the problem-solving techniques defined, the same model can be applied to the game trees in the sense of two-person games and to the state space from the single agent view. The two-person game tree is obtained from the intelligent tree model, as is the state space tree for agents.

Thus, a tree node m might be f(a1,a2,a3) & g(b1,b2), where *f* and *g* are Boolean functions of three and two variables, respectively, and ai's and bi's are Boolean-valued agents satisfying goal formulas for *f* and g. From the game tree viewpoint for what Shannon had estimated a complete tree carried to depth 6; three moves for each player; would already have one billion tip nodes.

Yet from an abstract mathematical viewpoint only, the game is a two; person game with perfect information. However, the two; person game view is not a mathematical model for any chess playing algorithm or machine. The real chess game, from the abstract viewpoint, might well be modeled as a multiagent game, being only a two; agent game with perfect information between mutually informal agents. A multiagent chess design or chess computing by any technique does not, in reality, provide perfect information in a way, which can be applied. The perfect information overall is a massive amount of data to be examined. There are thousands of move trees computed for searches coming close to being exhaustive. The multiagent chess paradigm, this author AAA 1997, is not based on an exhaustive two; person game with perfect information model. There is only minimal information for the multiagent plans across the board.

The multiagent multi-board model is a realization where the game is partitioned and correlated amongst agents and boards, with a cognitive anthropomorphism to human player's mind. There is an abstract two-person game model, but it does not apply to define a chess; playing machine. A tree game degree is the game state a tree is at with respect to a model truth assignment, e.g., to the parameters to the Boolean functions above. Let generic diagram or G-diagrams be diagrams definable by specific functions. We can then rank models based on game-tree satisfiability on a specific game tree degree. Thus, we have a model closest to a win when ranks higher on satisfiability.

Definition 5.1: A random diagram game tree is a game tree where assignments to variables are defined on a Boolean function on a specified random set.

We can then rank models based on game-tree satisfiability on a specific game tree degree.

Thus, we have a model closest to a win when ranks higher on satisfiability. Based on the above we can state basic theorems.

Proposition 3: A game tree node generic diagram for a player at a game tree node is satisfiable iff the subtree agent game trees situations have a compatible generic diagram to the node.

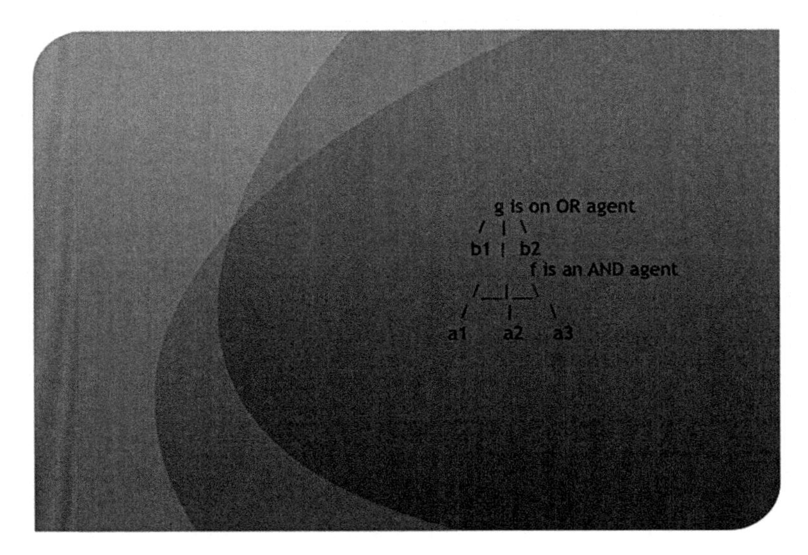

FIGURE 11.1 (See color insert.) Agent AND/OR tree.

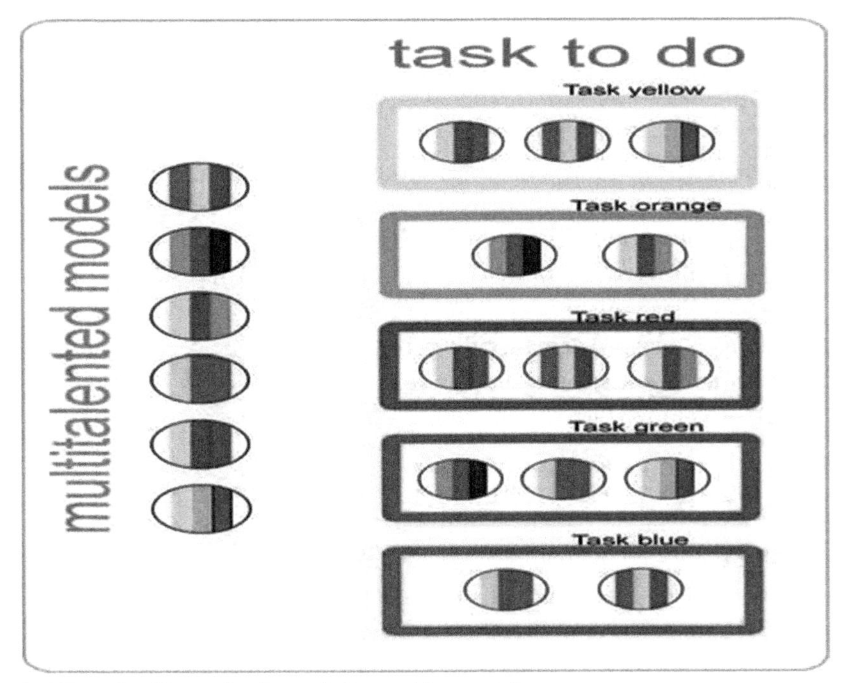

FIGURE 11.2 (See color insert.) Competitive model games.

Proof: Follows from game tress node satisfiability, the definitions in Section 11.2, definitions 3.2, 4.1, theorem 4.1, and the preceding statement culminating to definition 5.1. □

Proposition 4: A model has optimal rank iff the model satisfies every plan goal and has the minimum highest game tree node degree.

Proof: From proposition 3 the optimal rank is derived from compatible subtree satisfiability. Elementary embedding, c.f. Section 2.2, has upward compatibility on tree nodes as a premise. Applying Proposition 3, we can base diagram check on ordinal assignments to compute a countable conjunction, hence a minimal ordinal is reached for the model rank.

Definition 5.2: Let (M,a)c in C be defined such that M is a structure for a language L and each constant c in C has the interpretation a in M. The mapping c → ac is an assignment of C in M. We say that (M,a)c in C is a canonical model for a presentation P on language L, iff the assignment c→ a maps C onto M, i.e., M = (a: c in C).!

Theorem 5.1: There are computable models where optimal ranks can be determined.

Proof: Applying definition 5.2 from set model theory, with generic model diagram function, there are canonical model diagram computable functions, rendering computability on canonical initial tree model computing with known generic diagram functions. From the preceding propositions on model diagram compatibility on game trees, there are models considered for optimal rank computability. Amongst such models, there are models that are satisfiable with the minimality condition determined from a compatibility ordering implicit with model diagrams. A premise for having elementary embeddings is upward compatibility on tree nodes. One can diagram check the ordinal assignments in embeddings to compute a countable conjunction, hence an ordinal for the optimal rank is determined.

Theorem 5.2: Ranking Models, in general, is NP-Complete.

Proof: Ranking models are by definition based on game-tree satisfiability on a specific game tree degree. A model closest to a win when ranks higher on satisfiability. Therefore, ranking entails satisfiability, abbreviated as SAT in concrete complexity, an NP-complete problem. Hence, ranking, in general, is NP-complete.

Based on computable models first author's publications 2005–2009 on we have nicer computability criteria. On initial model computability the authors over two decades publications, and, e.g., for a comprehensive volume Horst Reichel 1987 can be an example. Additional expositions were (Bergstra and Tucker, 1981) and (Tucker, 2005).

Theorem 5.3: Based on computable models with computable diagrams model compatibility is effectively computable.

Proof: By definition 3.3 two generic diagrams D1 and D2 on the same logical language L are compatible iff forever assignment to free variables to models on £, every model M for L, M $\models$ D1 iff M $\models$ D2. Applying definition 5.2 from set model theory, with generic model diagram function, there are canonical model diagram computable functions, rendering computability on canonical initial tree model computing with known generic diagram functions.

11.6 AREAS TO FURTHER EXPLORE

Areas to explore from here is (Merkel and Mihalovic, 2004) where there are basic techniques to construct Martin-Löf random and rec-random sets with certain additional properties: any given set X we construct a Martin-Löf random set from which X can be decoded effectively. By a variant of the basic construction, one can obtain a rec-random set that is weak truth-table constructible. We observe that there are Martin-Left random sets that are computably enumerable self-reducible. The model reducibility areas that were carried on at ASL (Nourani, 1999) might be applicable to more standard business computational applications.

Denrell examines strategy concerned with sustained interim profitability differences where is shown that sustained interfirm profitability differences may be very likely even if there are no a priori differences among firms. As a result of the phenomenon of long leads in random walks, even a random resource accumulation process is likely to produce persistent resource heterogeneity and sustained interfirm profitability differences. Models in which costs follow a random walk shows that such a process could produce evidence of substantial persistence of profitability. The results suggest that persistent profitability does not necessarily provide strong evidence for systematic a priori differences among firms. Random competitive models are characterizing comparison for realistic applications.

11.7 CONCLUDING COMMENTS

Applications to forecasting techniques in the authors decade papers can be considered with the above to the more realistic example for a model of competitive stock trading in which investors are heterogeneous in their information and private investment opportunities and rationally trade for both informational and noninformational motives. For example, that volume is positively correlated with absolute changes in prices and dividends. However, that informational trading and random trading models lead to different dynamic relations between trading volume and stock returns. Competitive minimax approach to robust estimation of random parameters is another application area. So is competitive bidding (Rothkpf, 1968).

KEYWORDS

- **competitive model computing**
- **data analytics**
- **game degrees**
- **model compatibility**
- **model rank computability**
- **random model diagrams**
- **random sets**

REFERENCES

Berenbrink, P., & Schulte, O., (2010). Evolutionary equilibrium in Bayesian routing games: Specialization and niche formation. *Theor. Comput. Sci., 411*(7–9), 1054–1074.

Bergstra, J., & Tucker, J., (1981). Algebraic specifications of computable and semicursive approaches to abstract data types and the existence of recursive models. In: Goos and Hartmanis, (eds.), *LNCS, 118*, ISBN 3–540–10856–4

Chang, C. C., & Keisler, H. J., (1973). *Model Theory* (3rd edn.), *Studies in Logic and the Foundations of Mathematics*, Hardcover ISBN-13: 978-0444880543 ISBN-10: 0444880542.

Cressie, N., & Laslett, G. M., (1987). Random set theory and problems of modeling. *SIAM Review, 29*, 557–574.

Denrell, J., (2004). Random walks and sustained competitive. *Advantage Management Science, 50*(7), 922–934.

Gale, D., & Stewart, F. M., (1953). "Infinite games with perfect information." In: *Contributions to the Theory of Games, Annals of Mathematical Studies* (Vol. 28), Princeton.

Genesereth, M. R., & Nilsson, N. J., (1987). *Logical Foundations of Artificial Intelligence,* Morgan- Kaufmann.

Hintikka, J., (1963). *Knowledge and Belief,* Cornell University Press, Ithaca, N.Y.

Hoppe, T., (1992). " On the relationship between partial deduction and predictive reasoning." *Proc. 10th ECAI Conference* (pp. 154–158). August, Vienna, Austria, John Wiley and Sons.

Horst, R., (1987). Initial computability, algebraic specifications, and partial algebras (International series of monographs on computer science) Hardcover Oxford University Press. *International Series of Monographs on Computer Science (Book 2),* Oxford University Press; 1st edn., First Printing edition, Language: English ISBN–10: 0198538065.

Kendall, D. G., (1974). Foundations of a theory of random sets. In: Harding, E. F., & Kendall, D. G., (eds.), *Stochastic Geometry* (pp. 322–376). John Wiley, New York.

Kleene, S., (1956). *Introduction to Metamathematics.*

Kripke, S. A., (1963). "Semantical analysis of modal logic." Zeitschrift fuer Mathematische Logik und Grundlagen der Mathematik, *9*, 67–69.

Martin-Löf, (1966). "The definition of random sequences." *Information and Control, 9*(6), 602–619.

Nguyen, H. T., (2006). *An Introduction to Random Sets.* Chapman and Hall/CRC Press, Boca Raton, Florida.

Nilsson, N. J., (1969)."Searching, problem-solving, and game- playing trees for minimal cost solutions." In: Morell, A. J., (ed.), *IFIP 1968* (Vol. 2, pp. 1556–1562). Amsterdam, North-Holland.

Nilsson, N. J., (1971). *Problem Solving Methods in Artificial Intelligence,*" New York, McGowan-Hill.

Nourani, C. F., & Hoppe, T., (1993). "*GF Diagrams, and Free Proof Trees.*" Technical report, Abstract presented at the Berlin logic colloquium. Universitat Potsdam, Mathematics Department.

Nourani, C. F., & Oliver, S., (2013). "*Multiagent Decision Trees, Competitive Models, and Goal Satisfiability,*" DICTAP, Ostrava, Czeck Republic.

Nourani, C. F., & Oliver, S., (2014). *Multiagent Games, Competitive Models, and Descriptive Computing.* SFU, Vancouver, Canada, Preliminary Draft outline, New SFU computational logic lab, Technical report, Vancouver, Canada. Euromentt, Prague, Czeck Republic.

Nourani, C. F., & Schulte, O., (2013). "*Competitive Models, Descriptive Computing, and Nash Games,*" by AMS, Oxford, Mississipi, Reference: 1087–91–76.

Nourani, C. F., (1984). "Equational intensity, initial models, and reasoning in AI: A conceptual overview." *Proc. Sixth European AI Conference,* Pisa, Italy, North-Holland.

Nourani, C. F., (1994a). "A theory for programming with intelligent syntax trees and intelligent decisive agents- preliminary report." *11th European Conference AI.* ECAI Workshop on DAI Applications to Decision Theory, Amsterdam.

Nourani, C. F., (1994b). "Towards computational epistemology: A forward." *Proc. Summer Logic Colloquium*, Clermont-Ferrand, France. Memphis.

Nourani, C. F., (1995). "*Free Proof Trees and Model-theoretic Planning.*" Automated Reasoning AISB, England.

Nourani, C. F., (1996). "*Computability, KR, and Reducibility.* ASL, Toronto.

Nourani, C. F., (1999). *"Infinitary Multiplayer Games," Summer Logic Colloquium*. Utrecht.
Nourani, C. F., (2000). *"Descriptive Definability: A Model Computing Perspective on the Tableaux."* CSIT.: Ufa, Russia,msu.jurinfor.ru/CSIT2000.
Nourani, C. F., (2003). *"KR, Data Modeling, DB, and KB Predictive Scheme Completion."* A version published at International Systems and Cybernetics, Florida.
Nourani, C. F., (2005a). *Fragment Consistent Algebraic Models*, Brief Presentation, Categories, Oktoberfest, Mathematics Department, Ottawa, Canada.
Nourani, C. F., (2005b). *Functorial Model Computing, FMCS, Mathematics Department*, University of British Columbia Vancouver, Canada, www.pims.math.ca/science/2005/05fmcs.
Nourani, C. F., (2008a). *Business Planning and Cross-Organizational Models*, http://mc.informatik.uni -hamburg.de/konferenzbaende/mc2005/workshops/WS4_B2.pdf
Nourani, C. F., (2008b). *Fragment Consistent Kleene Models, Fragment Topologies, and Positive Process Algebras*, http://www.lix.polytechnique.fr, ATMS, France.
Nourani, C. F., (2009a). A descriptive computing, information forum, Leipzig, Germany. SIWN2009 Program, 2009. The Foresight Academy of Technology Press. *International Transactions on Systems Science and Applications* (Vol. 5, No. 1, pp. 60–69).
Nourani, C. F., (2009b). *Competitive Models, Game Tree Degrees, and Projective Geometry on Random Sets – A Preliminary*, ASL, Wisconsin.
Patel-Schneider, P. F., (1990). A decidable first-order logic for knowledge representation. *Journal of Automated Reasoning, 6*, pp. 361–388. A preliminary version published as AI Technical report number 45, Schlumberger Palo Alto Research.
Richter, M. M., (1978). *Logikkalkule, Teubner Studeinbucher,* Tubner, Leibzig.
Rothkopf, M. H., (1969). A model of rational competitive bidding. *Management Science*, pubsonline.informs.org.
Schulte, O., & Delgrande, J., (2004), Representing von Neumann-Morgenstern games in the situation calculus. *Annals of Mathematics and Artificial Intelligence, 42*(1–3), 73–101. (Special issue on multiagent systems and computational logic). A shorter version of this paper appeared in the 2002 AAAI, Workshop on Decision and Game Theory.
Schulte, O., (2003). Iterated backward inference: An algorithm for proper rationalizability. *Proceedings of TARK IX (Theoretical Aspects of Reasoning About Knowledge),* Bloomington, Indiana, pp. 15–28. ACM, New York. Expanded version with full proofs.
Slalom Tree Computing. A Computing Theory for Artificial Intelligence, (1994). (Revised December, 1994), A.I. Communication. *The European AI Journal*, *9*(4), IOS Press, Amsterdam.
Tucker, J. V., & Zucker, J. I. Computable total functions on metric algebras, universal algebraic specifications and dynamical systems, *The Journal of Logic and Algebraic Programming.*
Wolfgang, M., & Nenad, M., (2004). On the construction of effectively random sets. Published by: Association for Symbolic Logic. *The Journal of Symbolic Logic*, *69*(3), 862–878.
Zadeh, L. A., (1978). Fuzzy sets as a basis for a theory of possibility. *Fuzzy Sets and Systems, 1*, 3–28.

CHAPTER 12

LIVE GRAPHICAL COMPUTING LOGIC, VIRTUAL TREES, AND VISUALIZATION

CYRUS F. NOURANI

Acdmkrd-AI TU Berlin, Germany,
E-mail: cyrusfn@alum.mit.edu

ABSTRACT

The chapter is a preliminary overview of a virtual tree computing logic with applications to Virtual Reality. Meta-contextual logic is combined with Morph Gentzen, a new computing logic the author presented in 1997. It has applications towards Virtual Reality (VR) computing since the trees on the languages can carry visual structures via functions. Designated functions define agents, as specific function symbols, to represent languages with only abstract definition known at syntax. The languages are called "Intelligent" in the sense that there are designated function symbols on the syntax with explicit signature annotations, for example having to have 1–1 functions only on a sub signature. Generic diagrams for models are applied for a second order lift from visual context. The techniques allowed us to define a computational linguistics and model theory for "intelligent languages," i.e., agent augmented languages. Agent JAVA is applied to compute specific transformations for realistic designs.

12.1 INTRODUCTION

Starting with basic AI agent computing multimedia AI systems are presented necessitating aesthetics with world models. Multimedia context computing and context abstraction are introduced leading into visual object perception dynamics. Intelligent multimedia hybrid pictures and hybrid morphing is presented. Next, a high-level multimedia language is defined and its applications outlined. OOP and IOOP- Intelligent OOP are introduced with visual objects leading onto Virtual Visual computing with JAVA Agent computing with mobile visual objects conclude the chapter. The interest of AI for the past e25 years in concepts and theories of logic, computational linguistics, cognition, and vision was motivated by the urge to bridge the gap from logic, metaphysics, philosophy, and linguistics to computational theories of AI and practical AI systems. Practical artificial intelligence, AI, is concerned with the concepts and practice of building intelligent systems. Intelligent Multimedia techniques and paradigms are being defined whereby AI is applied to all facets of the enterprise starting from the basic multimedia programming stated in Nourani (2000). The VL computing paradigms and interfaces as presented at Burnett and Baker (1994) and Ambler et al. (1998) are paradigms which can be carried onto our design paradigms via IMPhora (Nournai, 2000) and virtual JAVA.

There are new computing techniques, languages, and a new deductive system with models. A new computing area is defined by Artificial Intelligence principles for multimedia. It is not a reinvention of AI, however, an exponential degree of computational expressibility is added when pictorial multimedia knowledge sources are augmented with agents and a complementing deductive system. Such computability is possible with the MIM and Morph Gentzen logic introduced.

12.2 AGENTS

The term "agent" has been applied to refer to AI constructs that enable computation on behalf of an AI activity. It also refers to computations that take place in an autonomous and continuous fashion, while considered a high-level activity, in the sense that its definition is software and hardware implementation, independent (Nourani, 1991a). For example, in mission planning (Genesereth-Nilsson, 1987; Nourani, 1991b) or space exploration, an agent might be assigned by a designed flight system (Nourani, 1991a) to compute the next docking time and location, with a known orbiting spacecraft. Intelligent agents are software entities that assist people and act on their behalf. Intelligent agents can automate the retrieval and processing of information. *Software agents* are specific agents designed to carry out specified tasks and define software functionality (e.g., Nourani

1995a, 1999a). Most agents defined by our examples are software agents. In the space applications we have had, of course, hard are functionality is specified for chip agents. Hence there are computing models with which agents might be applied the software counterpart to transistors and microchips. The example depicted by Figure 12.1 is from our double vision (Nourani, 1995c) and spatial computing projects. The visual field is represented by visual objects connected with agents carrying information amongst objects about the field and carried onto intelligent trees for computation. Intelligent trees compute the spatial field information with the diagram functions. The trees defined have function names corresponding to computing agents. The computing agent functions have a specified module defining their functionality.

12.3 AESTHETICS AND WORLD MODELS

According to Kant human knowledge is limited to appearances, whereas things in themselves are "noumena" i.e., thinkable but not actually knowable. Kant termed the doctrine Transcendental Idealism. Given the idealism is the possibility of synthesizing a priori knowledge to possible description and experience is easily explainable, since each object must necessarily conform to the conditions under which they can become objects for us. It assumes the human mind possesses such condition and demonstrating it is Transcendental Aesthetics. The computing import is since explored by the author's projects. To carry on with intelligent multimedia we have to be conscious as to what is the basic KR necessary to describe the knowledge and its significance for perception. KR with generic diagrams for models (Nourani, 1996d, 1996e) and applications define computable models and relevant world reasoning. G-diagrams are diagrams defined from a minimal set of function symbols that can inductively define a model. G-diagrams are applied to relevance reasoning by model- localized representations, and a minimal efficient computable way to represent relevant knowledge for localized AI worlds. We show how computable AI world knowledge is representable. Basic application areas to start with as examples are designing predefined visual scenes with diagram composition and the combination of scene dynamics. The second application area is based on AI planning (Fikes-Nilsson, 1971; Nourani, 1991). Reasoning and planning can be applied to define scene dynamics based on scene descriptions and compatibility relations. The project allows us to predict scene dynamics. We apply our recent Intelligent Language paradigm and intelligent visual computing paradigms to define the IM multiagent multimedia computing paradigm. By applying KR to define relevant worlds, personality parameters, combined with context compatibility and scene dynamics can be predicted. The role of context in KR and natural language systems, particularly in the process of reasoning, for example, is related to diagram functions defining relevant world knowledge for a particular context. The relevant world functions can proliferate the axioms and the relevant sentences for reasoning for a context. A formal computable theory can be defined based on the functions defining computable models for a context (Nourani, 1996d, 1997b). Abstract computational linguistics with intelligent syntax, model theory and categories is presented in brief. Designated functions define agents, as in artificial intelligence agents, or represent languages with only abstract definition known at syntax. For example, a function Fi can be agent corresponding to a language Li. Li can, in turn, involve agent functions amongst its vocabulary. This context might be defined by Li. Generic diagrams for models are defined as yet a second order lift from context. The techniques to be presented have allowed us to define a computational linguistics and model theory for intelligent languages. Models for the languages are defined by our techniques in (Nourani, 1995b, 1996f). KR and its relation to context abstraction are defined in brief.

12.4 MORPH GENTZEN

The IM Morphed Computing Logic Logics for computing for multimedia are new projects with important computing applications since Nouran (1997). The basic principles are a mathematical logic where a Gentzen or natural deduction systems are defined by taking arbitrary structures coded by diagram functions. The techniques can be applied to arbitrary topological structures. Thus, we define a syntactic morphing to be a technique by which infinitary definable structures are homomorphically mapped via their defining functions to new structures. The deduction rules are a Gentzen system augmented by two rules: morphing, and trans-morphing. The Morph Rule – A structure defined by the functional n-tuple <f1,…,fn> can be Morphed to a structures definable by the functional n-tuple <h(f1),…,h(fn)>, provided h is a homomorphism of abstract signature structures (Nourani, 1996). The TransMorph Rules – A set of rules whereby combining structures A1,…,An defines an Event {A1,A2,…,An} with a consequent structure B.

Thus, the combination is an impetus event. The deductive theory is a Gentzen system in which structures named by parameterized functions; augmented by the morph and trans-morph rules. The structures we apply the Morph logic to are definable by positive diagrams. The idea is to do it at abstract models syntax trees without specifics for the shapes and topologies applied. We start with $L\omega 1,\omega$, and further on might apply well-behaved infinitary languages.

12.4.1 THE MATHEMATICAL BASIS

12.4.1.1 AGENT AUGMENTED LANGUAGES

By an intelligent language, we intend a language with syntactic constructs that allow function symbols and corresponding objects, such that the function symbols are implemented by computing agents in the sense defined by Nourani (1997). A set of function symbols in the language, referred to by Agent Function Set, is a set of function symbols modeled in the computing world by Agents. The objects, messages passing actions, and implementing agents are defined by syntactic constructs, with agents appearing as functions, expressed by an abstract language that is capable of specifying modules, agents, and their communications. We have to put this together with syntactic constructs that could run on the tree computing theories. Sentential logic is the standard formal language applied when defining basic models. The language is a set of sentence symbol closed by finite application of negation and conjunction to sentence symbols. Once quantifier logical symbols are added to the language, the language of first-order logic can be defined. A Model for is a structure with a set A. There are structures defined for such that for each constant symbol in the language there corresponds a constant in A. For each function symbol in the language there is a function defined on A; and for each relation symbol in the language there is a relation defined on A. For the algebraic theories we are defining for intelligent tree computing in the forthcoming sections the language is defined from signatures as in the logical language is the language of many-sorted equational logic. The signature defines the language by specifying the function symbols' rarities. The model is a structure defined on a many-sorted algebra consisting of S-indexed sets for S a set of sorts. By an intelligent language, we intend a language with syntactic constructs that allow function symbols and corresponding objects, such that the function symbols are implemented by computing agents. A set of function symbols in the language, referred to by AF, is the set modeled in the computing world by AI Agents with across AND/OR overboard capability. Thus, the language defined by the signature has designated function symbols called AF. The AF function symbols define signatures which have specific message paths defined for carrying context around an otherwise context-free abstract syntax. A set of function symbols in the language, referred to by AF, are agents with nontrivial capability. The boards, message passing actions, and implementing agents are defined by syntactic constructs, with agents appearing as functions. The computation is expressed by an abstract language that is capable of specifying modules, agents, and their communications. Since the function symbols appearing might be invented by an activated agent without being defined in advance, intelligent Syntax allows us to program with nondeterministic syntax. The parsing problems are quite challenging. Trees connect by message sequences hence carry parsing sequences with them. Thus, the present computational linguistics theory is a start to programming with VAS and Nondeterministic Syntax.

12.4.1.2 AGENT LANGUAGES, VISUAL VIRTUAL TREES, AND CANONICAL MODELS

Linguistics knowledge representation and its relation to context abstraction are defined in brief. Nourani (e.g., Nourani, 1999a) has put forth new visual computing techniques for intelligent multimedia context abstraction with linguistics components as indicated in Section 11.5.2. In the present chapter, we also instantiate proof tree leaves with free Skolemized trees. Thus, virtual trees, at times like intelligent trees, are substituted for the leaves. By a virtual tree we mean a term made up of constant symbols and named but not always prespecified Skolem function terms. In virtual planning with G-diagrams that part of the plan that involves free Skolemized trees is carried along with the proof tree for a plan goal. We can apply predictive model diagram KR (Nourani, 2000) to compute queries and discover data knowledge from observed data and visual object images keyed with diagram functions. Model-based computing (Nourani, 1998c) can be applied to automated data and knowledge engineering with keyed diagrams. Specific computations can be carried out with predictive diagrams (Nourani, 1995a). For cognition, planning, and learning the robot's mind, a diagram grid can define the state. The starting space applicable project was meant for an autonomous robot's space journeys. The designs in the author's chapters are ways for a robot to update its mental state based on what it encountered on its path. That which the robot believes can be defined on a diagram grid. The degree to which a robot believes something is on the grid. It can get strengthened or weakened as a function of what the robot learns as progress is brought on. Robot's Mind State: The array grid entries are pointing to things to remember and the degree the robot believes them. The grid model is a way to encode the world with the model diagram functions.

12.4.1.2.1 Computing on Trees

In order to present some motivation for the methods proposed certain model-theoretic concepts are reviewed and some new techniques are presented. The Henkin style proof for Godel's completeness theorem is implemented by defining a model directly from the syntax of theories. A model is defined by putting terms that are probably equal into equivalence classes, then defining a canonical structure on the equivalence classes. The computing enterprise requires more general techniques of model construction and extension, since it has to accommodate dynamically changing world descriptions and theories. The models to be defined are for complex computing phenomena, for which we define generalized diagrams. The author's techniques for model building, e.g., Nourani (1991, 1996) as applied to the problem of AI reasoning allows us to build and extend models through diagrams. This required us to focus attention on generic model diagrams. The author introduced generic diagrams in the 1980's to build models with a minimal family of generalized Skolem functions. The minimal set of function symbols is the set with which a model can be inductively defined. The models are standard and computable. The G-diagram methods allowed us to formulate AI world descriptions, theories, and models in a minimal computable manner. Thus, models and proofs for AI and computing problems can be characterized by models computable by a set of functions. It allows us to program with objects and functions "running" on G-diagrams. To allude to our AI planning techniques as an example, the planning process at each stage can make use of generic diagrams G-diagrams with free Skolemized trees, by taking the free interpretation, as tree-rewrite computations for the possible proof trees that correspond to each goal satisfiability. Suppose there are some basic Skolem functions f1,…,fn that define a G-diagram. During planning or proof tree generation a set of Skolem functions g1,…,gn could be introduced (Nourani,

1995; Nourani-Hoppe, 1995). While defining such free proof trees, a set of congruence relations relates the g's to the f's. The proofs can make use of the tree congruence relations, or be carried out by tree rewriting.

The computing and reasoning enterprise requires more general techniques of model construction and extension, since it has to accommodate dynamically changing world descriptions and theories. The techniques in the author's projects for model building as applied to the problem of AI reasoning allows us to build and extend models through diagrams. A technical example of algebraic models defined from syntax had appeared in defining initial algebras (ADJ, 1977) for equational theories of data types (Nourani, 1980, 1996). In such direction for computing models of equational theories of computing, problems are presented by a pair (Σ,E), where Σ is a signature (of many sorts, for a sort set S) and E a set of Σ-equations. Let $T<\Sigma>$ be the free tree word algebra of signature Σ. The quotient of $T<\Sigma>$, the word algebra of signature Σ, with respect to the Σ-congruence relation generated by E, will be denoted by $T<\Sigma,E>$, or $T<P>$ for presentation P. $T<P>$ is the "initial" model of the presentation P. The Σ-congruence relation will be denoted by $\equiv P$. One representation of T(P) which is nice in practice consists of an algebra of the canonical representations of the congruence classes, abbreviated by Σ-CTA. It is a special case of generalized standard models the author had defined (Nourani, 1996 for newer examples). Some definitions are applied to the chapters that allow us to define standard models of theories that are Σ-CTA's. The standard models are significant for tree computational theories that the author had presented. Generic diagrams are applied to define canonical standard models in the same sense as a set theory. These definitions are basic to sets and in defining induction for abstract recursion and inductive definitions. We had put forth variants of it with axiomatizations in our papers. The definitions were put forth by the first author for the computability with initial models. The canonical models are applied to multiagent computing during the last several years by the author.

Definition 3.3: We say that a signature Σ is intelligent iff it has intelligent function symbols. We say that a language has intelligent syntax if the syntax is defined on an intelligent signature. D To define a specific mathematical linguistics basis for agent augmented languages intelligent languages were defined (Nourani, 1995d) as follows.

Definition 3.4: A language L is said to be an intelligent language iff L is defined from an intelligent syntax.

Agent augmented languages and signatures allow us to present computational theories with formulas on terms with intelligent function symbols.

12.4.1.2.2 Abstract Intelligent Syntax

It is essential to the formulation of computations on intelligent trees and the notion of congruence that we define tree intelligence content. A reason is that there could be a loss of tree intelligence content when tree rewriting because not all intelligent functions are required to be on mutual message exchanges. Theories are presented by axioms that define them and it is difficult to keep track of what equations not to apply when proving properties. What we have to define, however, is some computational formulation of intelligence content such that it applies to the present method of computability on trees. Once that formulation is presented, we could start decorating the trees with it and define computation on intelligent trees. It would be nice to view the problem from the standpoint of an example.

The examples of agent augmented languages we could present have <O,A,R> triples as control structures. The A's have operations that also consist of agent message passing. The functions in AFS are the agent functions capable of message passing. The O refers to the set of objects and R the relations defining the effect of A's on objects. Amongst the functions in AFS only some interact by message passing. What is worse the functions could affect objects in ways that affect the intellectual content of a tree. There you are: the tree congruence definition thus is more complex for agent augmented languages than those of ordinary syntax trees. Let us define tree intelligence content for the present formulation.

Definition 4.1: We say that a function *f* is a string function, iff there is no message passing or information exchange except onto the object that is at the range set for f, reading parameters visible at each object. Otherwise, *f* is said to be a splurge function. We refer to them by string and splurge functions when there is no ambiguity.

Remark: Nullary functions are string functions.

Definition 4.2: The tree intelligence degree (TID) is defined by induction on tree structures:

(i) the intellectual content of a constant symbol function *f* is f;
(ii) for a string function f, and tree f(t1,…,tn) the TID is defined by U TID (ti: : f), where (ti: : f) refers to a subtree of ti visible to f;

(iii) for a splurge function f, TID is defined by U TID (f: ti), where f: ti refers to the tree resulting from ti upon information exchange by f.

There are implicit mobile object computing principles at definition 4.2 for example, the concept of a subtree being visible to a function, and of course, agents. The theorem below formalizes these points. Thus, out of the forest of intelligent trees, there appears an information theoretic rewrite theorem.

Definition 4.3: We say that an equational theory T of signature IΣ is an intelligent IΣ theory iff for every proof step involving tree rewriting, the TID is preserved. We state T<IST> |- t=t' when T is an IΣ theory.

Definition 4.4: We say that an equational theory T is intelligent, iff T has an intelligent signature IΣ, and axioms E, with IΣ its intelligent signature. A proof of t=t' in an intelligent equational theory T is a finite sequence b of IΣ -equations ending in t=t' such that if q=q' is in b, then either q=q' in E, or q=q' is derived from 0 or more previous equations in E by one application of the rules of inference. Write T <IST>|- t=t' for "TP proves t=t' by intelligent algebraic subtree replacement system."

By definition of such theories proofs only allow tree rewrites that preserve TID across a rule. These definitions have been applied to prove the theorems, set up the foundations for intelligent tree rewriting and intelligent tree computation. Algebraic subtree replacement systems applying Lawvere algebraic theories were treated in the author's doctoral dissertation years ago. The tree-replacement systems are not directly deployed here. The techniques are a basis to have a direct model construction with morph rewriting. Thus, the essence of intelligent trees will not be lost while agent tree computing. Next, we define a computing agent function's intelligence content from the above definition. That is a matter of the model of computation applied rather than a definition inherent to intelligent syntax. Let us present it as a function of intelligent syntax only, because we are to stay with abstract models and build models from abstract syntax. The definition depends on the properties of intelligent trees defined as follows and Nourani (1996). Viewing the methods of computation on trees presented in the sections above we define intelligent trees here.

Definition 4.5: A tree defined from an arbitrary signature Σ is intelligent iff there is at least one function symbol *g* in Σ such that *g* is a member of the set of intelligent functions AFS, and *g* is a function symbol that appears on the tree.

Definition 4.6: We define an intelligent Σ-equation, abbreviate by IΣ-equation, to be a Σ-equation on intelligent Σ-terms. An IΣ congruence is a Σ -congruence with the following conditions:

(i) the congruence preserves IΣ equations;
(ii) the congruence preserves computing agents intelligence content of Σ-trees.

Canonical models are definable with canonical sets C on the carriers with <function,base-set> pair by recursions such that C with a set of tree rewrite rules R represents T<IΣ,~R>, where ~R is the set R of axioms for P viewed as IΣ-rewrite rules.

Definition 4.7: Let Σ be an intelligent signature. Then, a canonical term IΣ-algebra (IΣ-CTA) is a Σ-algebra C such that

(i) |C| is a subset of T<Σ> as S-indexed families;
(ii) gt1.tn in C implies ti's are in C, and gC (t1,…,tn) = gt1.tn, where gC refer to the operation in algebra C corresponding to the function symbol g.

For constant symbols (2) must hold as well, with gC = g.

(iii) gt1.tn in T<AFS> implies ti's in C; and gC(t1,…,tn) = gt1.tn; for constant symbols it must hold as gC=g.

Lemma 4.1: Let C be an IΣ-algebra. Let P = (Σ,E) be a presentation. Then, C is Σ-isomorphic to T<P>, iff

(i) C satisfies E;
(ii) gC (t1,…,tn) ≡P g.t1.tn
(iii) gC(t1,…,tn) ≡P gt1.tn, with gt1.tn in T<AFS> whenever ti's are in T<AFS> and gC is in AFS.

Note: (ii and iii) must also hold for constants with g.C = g; ≡ refers to the IΣ-congruence generated by E.

Proof: Follows from definitions and ADJ (1976). (c.f. Nourani, 1996). Nourani (1996) states sufficient conditions for constructibility of an initial model for an I Σ equational presentation from the above proposition. It is the mathematical justification for the proposition that initial models with intelligent signature can be automatically implemented (constructed) by algebraic subtree replacement systems. The normal forms are defined by a minimal set of functions that are Skolem functions or type constructors. Thus, we have the following Canonical Intelligent Model Theorems. The theorems provide conditions for automatic implementation by intelligent tree rewriting to initial models for programming with objects. The intelligent languages basis (Nourani, 1995d) defines intelligent context-free grammars as follows.

Definition 4.9: A language L is intelligent context-free, abbreviated by ICF, iff L is intelligent and there is a context-free grammar defining L. A preliminary parsing theory might be defined once we observe the correspondence between String Functions and context. Let us define string intelligent functions.

Definition 4.10: A language is String Intelligent iff it is an intelligent language and all agent functions in the language are 1–1 functions. The following starts at ICF theory is from ECAI (Nourani, 1994.)

The following is proved as a theorem: Let L be a context-free language with signature Σ. Let L* be a string intelligent language extending L such that L* has the same signature as L, except for string agent function symbols augmenting L's signature. Then, L* is ICF. The proof outline is that there is an initial algebra TΣ that is defined as directly from a context-free grammar's (ADJ, 1973) productions. The string agent functions on 1–1, therefore, there is an embedding homomorphism preserving the T Σ context-free trees.

12.4.2 VISUAL AND VIRTUAL TREES

Linguistics knowledge representation and its relation to context abstraction are defined in brief. Nourani (e.g., Nourani, 1999a) has put forth new visual computing techniques for intelligent multimedia context abstraction with linguistics components. In the present chapter, we also instantiate proof tree leaves with free Skolemized trees. Thus, virtual trees, at times like intelligent trees, are substituted for the leaves. By a virtual tree we mean a term made up of constant symbols and named but not always prespecified Skolem function terms. In virtual planning with G-diagrams that part of the plan that involves free Skolemized trees is carried along with the proof tree for a plan goal. We can apply predictive diagram KR (knowledge representation) to compute queries and discover data knowledge from observed data and visual object images keyed with diagram functions. Model-based computing (Nourani,

1998c) can be applied to automated data and knowledge engineering with keyed diagrams. Specific computations can be carried out with predictive diagrams (Nourani, 1995a). For cognition, planning, and learning the robot's mind, a diagram grid can define the state. The starting space applicable project was meant for an autonomous robot's space journeys. The designs in the author's s are ways for a robot to update its mental state based on what it encountered on its path. That which the robot believes can be defined on a diagram grid. The degree to which a robot believes something is on the grid. It can get strengthened or weakened as a function of what the robot learns as progress is brought on. Robot's Mind State: The array grid entries are pointing to things to remember and the degree the robot believes them. The grid model is a way to encode the world with the model diagram functions.

Canonical Models from models to set theory had been stated for arbitrary structures as follows.

Generic diagrams were present in the preceding chapters (e.g., Chapters 7 and 10) allow us to define virtual tree canonical models with specific functions. For free Skolemization what comes to mind is generic model expansion. It is relevant to our methods of modeling, planning, and reasoning. A virtual tree, or virtual proof tree, is a proof tree that is constructed with agent languages with free Skolem functions.

Definition 4.12: A G-diagram for a structure M is a diagram such that the there is a proper diagram definition with specific function symbols, for example, Σ1 Skolem functions.

Theorem 2: For the virtual proof trees defined as a goal formula from the G-diagram there is an initial model satisfying the goal formulas. It is the initial model definable by the G-diagram.

Proof: In planning with GF-diagrams plan trees involving free Skolemized trees is carried along with the proof tree for a plan goal.

The idea is that if the free proof tree is constructed then the plan has a model in which the goals are satisfied. There is an analogy to SLD proofs. We can view on the one hand, SLD resolution type proofs on ground terms, where we go from p(0) to p(f(c)); or form p(f(c)) to p(f(g(c)). Whereas, while doing proofs with free Skolemized trees we are facing proofs of the form p(g(.)) proves p(f(g(.)) and generalizations to p(f(x)) proves For all x, p(f(x)). Since the proof trees are either proving plan goals for formulas defined on the GF-diagram or are computing with Skolem functions defining the GF-diagram, the model defined by the GF-diagram applies and it is initial for the proofs.

For free Skolemization what comes to our logician mind's side of things is Generic model expansion. It is relevant to our methods of modeling, planning, and reasoning. A free proof tree is a proof tree that is constructed with free Skolem functions and GFS from a GF-diagram. The idea is that if the free proof tree is constructed for a goal formula, the GF-diagram defines a model satisfying the goal formula satisfied. The model is the initial model of the AI world for which the free Skolemized trees were constructed. Thus, we have transformed the model-theoretic problems of computing to that of defining computing with Generalized Diagrams. There are recent papers since 1992 (ADJ, 1973; Nourani Hoppe, 1994), where we have presented this theory and technique. Partial deductions in the present approach correspond to proof trees that have free Skolemized trees in their representation. These concepts will be developed in forthcoming papers by this author. In the present approach, the free proof tree technique, as we shall further define, leaves could be virtual, where virtual leaves are free Skolemized trees.

For plans with free Skolmized trees, we can apply the Hilbert epsilon technique to define computing models. What applying Hilbert's epsilon implies is that there is a model M for the set of formulas such that we can take an existentially quantified formula w[X] and have it instantiated by a Skolem function that can answer the membership question to the model. Whether or not Hilbert had intended it this way or not is not relevant at present. The issue in our approach, however, is that we are not so much concerned with existentially quantified formulas. We start with some Skolem functions to define Initial models. We have planning applications with VR in which there is goal formula to be satisfied with perhaps existential quantifiers. Since we are interested in model-theoretic techniques for handling proofs with the method of free proof trees we propose the following model-theoretic view, which we refer to by the Hilbert Model Theorem for Skolemized virtual tree computing.

Theorem 3: The Hilbert's epsilon technique implies there is a model M for the set of formulas such that we can take an existentially quantified formula w[X] and have it instantiated by a Skolem function which can answer the satisfiability question for the model.

Proof: We start with some Skolem functions to define the models. The equivalence defined by Hilbert's epsilon function is instantiated by the Skolem functions defining GF-diagrams. Therefore, the plans have Hilbert models for which the Skolem functions implicitly define membership to sets, defining a model. While doing proofs with free Skolemized trees we are facing proofs of the form p(g(.)) proves p(f(g(.)) and generalizations to p(f(x)) proves for all x, p(f(x)). Since the proof trees are either proving plan goals for formulas defined on the G-diagram or are computing with Skolem functions defining the GF-diagram, the model defined by the GF-diagram applies for the proofs.

12.4.3 POSITIVE MORPHISMS AND MODELS

From model-theoretic foundations we have the following:

Definition 4.13: A formula is said to be positive iff it is built from atomic formulas using only the connectives &, v, and the quantifiers∀, ∃.

Definition 4.14: A formula ϕ (x1,x2,…,xn) is preserved under homomorphisms iff for any homomorphisms *f* of a model A onto a model B and all a1,…,an in A if A |= [a1,…,an] B |= [fa1,…,fan].

Theorem 4: A consistent theory is preserved under homomorphisms iff T has a set of positive axioms. □

12.4.3.1 POSITIVE MORPHISMS AND MODELS

Definition 4.15: A formula is said to be positive iff it is built from atomic formulas using only the connectives &, v, and the quantifiers∀, ∃.

Definition 4.16: A formula ϕ (x1,x2,…,xn) is preserved under homomorphisms iff for any homomorphisms *f* of a model A onto a model B and all a1,…,an in A if A |= [a1,…,an] B |= [fa1,…,fan].

Theorem 5: A consistent theory is preserved under homomorphisms iff T has a set of positive axioms.

Proof: Well-known theorem that is proved based on the basic homomorphic properties, elementary model embedding, and tower limit chains.

12.4.4 MORPH GENTZEN MODEL

As discussed earlier, the deduction rules of a Gentzen system are argmented by two rules. They are morphing, and trans-morphing. In the Morph rule, a structure is defined by the functional n-tuple <f1,…,fn> that can be morphed to a structures definable by the functional n-tuple <h(f1),…,h(fn)>, provided h is a homomorphism of abstract signature structures (Nourani, 1996). Whereas, in the transmorph rules, there is a set of rules whereby combining structures A1,...,An that defines an event {A1,A2,...,An} with a consequent structure B. Thus, the combination is an impetus event. In the Gentzen system (i.e., deductive theory) the structures are named by parameterized functions; augmented by the morph and trans-morph rules. The positive diagrams are definable by the structures that we apply to the morph logic. The idea is to do it at abstract models syntax trees without specifics for the shapes and topologies applied. We start with $L\omega 1,\omega$, and further on might apply well-behaved infinitary languages.

FIGURE 12.1 **(See color insert.)** Morphs and trans-morphs.

Theorem 6: Soundness and Completeness–Morph Gentzen Logic is sound and complete.

Proof: outline Plain Morph Gentzen Logical Completeness has two proofs:

(i) There is a conventional proof route whereby we start with the completeness theorem for ordinary Gentzen systems. Form it we can add on the morph rules and carry out a proof based on what the morph rules preserve on models. Again intricate models are designed with positive diagrams following Section 12.4.1.2. The Morph Rule A structure defined by the functional n-tuple <f1,...,fn>

can be Morphed to a structures definable by the functional n-tuple <h(f1),…,h(fn)>, provided h is a homomorphism of abstract signature structures. The TransMorph Rules: A set of rules whereby combining structures A1,…,An defines an Event {A1,A2,…,An} with a consequent structure B. The rules preserve Morph Gentzen completeness based on Section 12.4.

(ii) There is a direct proof which applies positive diagrams, and canonical models for L $\omega 1,\omega$ fragments as the authors publications, e.g., (Nourani, 1996) in mathematical logic. "Functorial Model Theory and Infinite Language Categories," ASL, SF, January 1995. Bulletin ASL, Vol.2, Number 4, December 1996.

The direct proof has to apply what the author had developed since 1980's on infinitary logic fragments with what might come close to intuitionistic logic. For example, Gödel (1932) intuitionistic propositional logic that has disjunction property and Gentzen (1935) proved the disjunction property for closed formulas of intuitionistic predicate logic. Kleene (1945, 1952) proved that intuitionistic first-order number theory and the related existence property: If xA(x) is a closed theorem, then for some closed term t, A(t) is a theorem. Intuitionistic systems were the predecessors to Beth's tableaus, Rasiowa, and Sikorski's topological models, and Kleene's recursive realizabilities. These areas are further explored on realizability and Beth tableaus in Nourani (2000, 2007) and onto Kripke's (1965) possible-world semantics (Nourani Scandinivina AI, 1991), with respect to which intuitionistic predicate logic is complete and consistent on classical model theory.

Proposition Morph Gentzen and agent augmented languages provide a sound and complete logical basis to VR.

12.5 CONCLUSION

The chapter presented a novel deductive system for graphical and visual computing with multiagent intelligent multimedia techniques. What might come closest to how the mathematical formalisms are applied to visual data is the proceedings on diagrammatic reasoning (Galsgow, et al., 1995). However, the techniques here have to specific relations to what had transpired there. Furthermore, that specific areas was not agent computing or intelligent multimedia at all. Rather, a static characterization on visual data. The techniques here had encamped a preliminary overview to context abstraction and meta-contextual logic with applications, abstract computational linguistics (Nourani, 1995) with intelligent syntax, model theory, and categories. Meta-contextual logic is combined with Morph Gentzen, a new computing logic the author presented in 1997, towards a Virtual Reality design languages and their computing logic and Haptic computing (Picard, 1999; Nourani, 2005, 2006).

KEYWORDS

- **morphs and trans-morphs**
- **multiagent intelligent multimedia techniques**
- **novel deductive system**

REFERENCES

Ambler, A. L., & Jennifer, L., (1997). "Public Programming in a Web World." In: *1998 IEEE Symposium on Visual Languages*, Halifax, Nova Scotia.

Anderson, M., (1994). "Reasoning with diagram sequences." In: *Proceedings of the Conference on Information-Oriented Approaches to Logic, Language, and Computation (Fourth Conference on Situation Theory and its Applications)*. Moraga, California.

Àngel, J. G., Jordi, R., & Ventura, V., (1999). A Strong Completeness Theorem for the Gentzen systems associated with finite algebras. *Journal Title: Journal of Applied Non-Classical Logics, Date, 9*(1).

Barwise, J., (1996). *Logical Reasoning with Diagrams*, Oxford University Press, (ISBN: 0–19–510427–7).

Berkeley, C. A., (1997). Preliminary edition. Nourani, C. F., "Intelligent Multimedia- New Computing Techniques and its Applications," CSIT'99, In: Freytag, & Wolfengagen, V., (eds.), *Proceedings of 1st International Workshop on Computer Science and Information Technologies*, Moscow, Russia. MEPhI Publishing 1999, ISBN 5–7262–0263–5.

Burnett, M. M., & Marla, J. B., (1994). A classification system for visual programming languages. *Journal of Visual Languages and Computing*, 287–300.

Ehrig, H., (1990). *Graph Grammars and Their Application to Computer Science. 4th International Workshop*, Bremen, M. H., & Ehrig, et al. (eds.), Springer Lecture Notes in. www.informatik.uni-mainz.de/~goettler/Goettlerpub.html.

Genesereth, M., & Nilsson, N. J., (1987). *Logical Foundations of Artificial Intelligence*, Morgan Kaufmann.

Gentzen, G, B., (1943). Unbewiesbarket von Anfangsfallen der trasnfininten Induktion in der reinen Zahlentheorie, *Math Ann., 119*, 140–161.

Glasgow, J. L., Narayanan, N. H., & Chandrasekaran, B., (1995). *Diagrammatic Reasoning: Cognitive and Computational Perspectives*. Boston, MA: MIT Press and Menlo Park, CA: AAAI Press.

Gödel, K., (1932). "Zum intuitionistischen aussagenkalkül," *Anzeiger der Akademie der Wissenschaftischen in Wien, 69*, 65–66.

Gödel, K., (1933). "Zur intuitionistischen Arithmetik und Zahlentheorie," *Ergebnisse Eines Mathematischen Kolloquiums 4*, 34–38.

Goguen, A. D. J., Thatcher, J. W., Wagner, E. G., & Wright, J. B., (1973). *A Junction Between Computer Science and Category Theory, (Parts I and II)*. IBM T.J. Watson Research Center, Yorktown Heights, N.Y. Research Report, RC4526.

Heyting, A., (1930). "Die formalen Regeln der intuitionistischen Logik," in three parts, Sitzungsber. preuss. *Akad. Wiss, 42–71,* 158–169.

Heyting, A., (1971). Intuitionism: An Introduction, Amsterdam: North-Holland (1956). Third Revised Edition. In: Nourani, C. F., (ed.), *Descriptive Definability: A Model Computing Perspective on the Tableaux*. CSIT 2000: Ufa, Russia, msu.jurinfor.ru/CSIT2000/.

Hirakawa, M., & Tadao, I., (1994). Visual languages studies – A perspective. In: *Software – Concepts and Tools* (pp. 61–67).

Keiselr, H. J., (1978). Fundamentals of model theory. *Handbook of Mathematical Logic, A2*, (Barwise, editor), North-Holland.

Kleene, S. A., (1951). *Introduction to Met Mathematics*. North-Holland.

Lambek, J., (1995). The mathematics of sentence structure. *American, Mathematical Monthly, 65*, 154–170.

Nourani, C. F., (1906–2006). *Positive Omitting Types and Fragment Consistency Models.* Celebrating Kurt Gödel's Centennial, Brazilian Logic Society and Association for Symbolic Logic. www.tecmf.inf.puc-rio.br/EBL06/Program.

Nourani, C. F., (1987). *Diagrams, Possible Worlds, and the Problem of Reasoning in Artificial Intelligence*, Logic Colloquium, Padova, Italy.

Nourani, C. F., (1994). *A Theory for Programming With Intelligent Syntax Trees and Intelligent Decisive Agents*, 11th ECAI–94, Workshop on Decision Theory for AI, Amsterdam.

Nourani, C. F., (1995). "*Free Proof Trees and Model-Theoretic Planning*, Automated Reasoning AISB, Sheffield, England.

Nourani, C. F., (1995). *Automatic Models From Syntax.* The XVth Scandinavia Conference of Linguistics, Oslo, Norway.

Nourani, C. F., (1995a). *Double Vision Computing, IAS–4, Intelligent Autonomous Systems.* Karlsruhe, Germany.

Nourani, C. F., (1995a). *Abstract Linguistics- A Brief Overview*, ICML, Mathematical Linguistics, Tarragona, Catalunia, Spain.

Nourani, C. F., (1995b). *Intelligent Syntax, Abstract Linguistics, and Agents*. ACL String.

Nourani, C. F., (1996a). *Slalom Tree Computing, 1994, AI Communications* (Vol. 9. No.4). IOS Press, Amsterdam.

Nourani, C. F., (1996b). *Autonomous Multiagent Double Vision Space Crafts*, AA99- Agent Autonomy Track, Seattle, WA.

Nourani, C. F., (1997). *"MIM Logik," Summer Logic Colloquium*, Prague.

Nourani, C. F., (1998b). *Intelligent Trees, Thought Models, And Intelligent Discovery.* Model-Based Reasoning in Scientific Discovery (MBR'98) Pavia, Italy.

Nourani, C. F., (1998c). "Visual computational linguistics and visual languages." *Proceedings 34th International Colloquium on Linguistics*. University of Mainz, Germany.

Nourani, C. F., (1998d). *Morph Gentzen, KR, and World Model Diagrams*. Automated Deductions and Geometry, ETH Zurich.

Nourani, C. F., (1999). *Visual Computational Linguistics.* Brief at the 34th Linguistics Colloquium, Mainz, Germany.

Nourani, C. F., (1999). *Intelligent Multimedia New Computing Techniques, Design Paradigms, and Applications*, http://www.treelesspress.com/.

Nourani, C. F., (1999–2000). *Descriptive Definability and the Beth Tableaux*. Winter Meeting of the Association for Symbolic Logic.

Nourani, C. F., (1999a). "Multiagent AI implementations an emerging software engineering trend," *Engineering Applications of AI, 12*, 37–42.

Nourani, C. F., (2000). *A Multiagent Intelligent Multimedia Visualization Language and Computing Paradigm.* VL2000. www.Interface.Track, Seattle.

Nourani, C. F., (2000). *Model Discovery and Knowledge Engineering with Predictive Logic*. Fukuoka, Japan, www.information-iii.org/conf/info2000.html.

Nourani, C. F., (2000a). "Versatile abstract syntax meta-contextual logic and VR computing 36th Lingustische Kolloquium, Austria." *Proceedings of the 35th Colloquium of Linguistics*. Aesthetics, CL, and Models. Friedrich-Schiller-Universität Jena Institut für Germanistische Sprachwissenschaft, Institut für Anglistik/Amerikanistik Fürstengraben 30, Ernst-Abbe-Platz 8The 37th Linguistic Colloquium Germany. Motto "Language and the modern media." http://www.uni-jena.de/fsu/anglistik/Lingcoll/.

Nourani, C. F., (2003). "Versatile abstract syntax meta-contextual logic and VR computing." 36th Lingustische kolloquium, *Austria Proceedings of the 35th Colloquium of Linguistics*, Innsbruck. Europa Der Sprachen: Sprachkopetenz-Mehrsprachigeit-Translation, TIEL II: Sprache und Kognition, Sonderdruc, Lew N. Zybatow (HRSG.)

Nourani, C. F., (2003). *A Sound and Complete Agent Logic Paradigm, 2000.*" Parts and abstract published at ASL, and FSS, AAAI symposium.

Nourani, C. F., (2004). *Intelligent Multimedia Computing Science Business Interfaces*. Wireless Computing, Databases, and DataMines, ISBN: 1–58883–037–3, www.asp.org/multimimedia.html.

Nourani, C. F., (2004). Versatile abstract syntax: Liberating programming from deterministic syntax. *The Third International Conference on Information*, Boissonade Tower 26F, Hosei University, Tokyo, Japan.

Nourani, C. F., (2006). *A Haptic Computing Logic – Agent Planning, Models, and Virtual Trees The Future of Learning* (Vol. 1). Affective and emotional aspects of human-computer interaction – game- based and innovative learning approaches, Edited by Maja Pivec ISBN: 978–1–58603–572–3. Affiliations Academic USA and the Scientific URL http://projektakdmkrd.tripod.com

Nourani, C. F., (2007). *Positive Realizability Morphisms and Tarski Models.*, Logic Colloquium '07 Wrocław, Poland.

Nourani, C. F., (1995d). *Intelligent Languages- A Preliminary Syntactic Theory, Mathematical Foundations of Computer Science, (1998).* 23rd International Symposium, Brno, Czech Republic, August The satellite workshop on Grammar systems. Silesian University, Faculty of Philosophy and Sciences, Institute of Computer Science.

Picard, R. W., (1999a). *Affective Computing for HCI, Proceedings of HCI,* Munich, Germany.

Wooldridge, M., & Jennings, N. R., (1995). Agent theories, architectures, and languages: A survey. *Intelligent Agents,* 51–92.

CHAPTER 13

INFERENCE TREES, COMPETITIVE MODELS, AND SPARSE BIG DATA HEURISTICS

CYRUS F. NOURANI and JOHANNES FÄHNDRICH

Berlin Institute of Technology, DAI Lab, Berlin, Germany

ABSTRACT

Novel predictive modeling analytics techniques with decision trees on competitive models with big data heuristics are presented in brief. Spanning trees are applied to focus on plan goal models that can be processed with splitting agent trees,

The chapter is based on agent plan computing where the interaction amongst heterogeneous computing resources is via objects, multiagent AI and agent intelligent languages. Modeling, objectives, and planning issues are examined at an agent planning. Further developing the techniques for model discovery and prediction planning first author developed from the preceding chapters we examine inference decision trees with application to big data areas. Based on the first authors past decade on the preceding chapters, agent computing techniques are applied to present precise decision strategies on multiplayer games with only perfect information between agent pairs on a vector spanning models. We brief on how sparse matrices enable efficient computability on big date heuristics. Predictive analytics on "big data" are presented with new admissibility criteria. Tree computing grammar algebras graphs are a basis for realizing tree goal planning.

13.1 INTRODUCTION

The areas explored in this chapter range from plan goal decision tree satisfiability with competitive business models to predictive analytics models that accomplish goals on a 3-tier business systems design models. Attention spanning trees are applied to focus on plan goal models that can be processed on a vector state machine coupled with a database pre-processor data mining interfaces (Nouran-Lauth-Pedersen, 2015). Modeling, objectives, and planning issues are examined to present precise decision strategies to satisfy such goals. Heuristics on predictive analytics are examined with brief applications to decision trees. Business intelligence and analytics (BI&A) and the related field of big data analytics have become increasingly important in both the academic and the business communities over the past two decades. Section outlines are as follows: Section 13.2 presents the basics on competitive goals and models. Agent AND/OR trees are applied as primitives on decision trees to be satisfied by competitive models. Section 13.3 briefs on goals, plans, and realizations with database and knowledge bases. We brief on a function key interface to the databases from Nourani (2001) TAIM database design. Competitive model goal satisfiability with model diagrams is glimpsed. Section 13.4 presents the applications to practical decision systems design with splitting agent decisions trees. The chapter concludes with heuristics for competitive models and goals comparing with Russell and Norvig (2002) from the first author's newer economics game decision tree bases.

13.2 MODELS AND GOALS

Practical systems are designed by modeling with information, rules, goals, strategies, and information onto the data and knowledge bases, where masses of data and their relationships and representations are stored, respectively. A novel basis to decision-theoretic planning with competitive models was presented in Nourani (2005) and Nourani-Schulte (2013) with classical and non-classical planning techniques from artificial intelligence, games, and decision trees, providing an agent expressive planning model. Planning with predictive model diagrams represented with keyed to knowledge bases is presented (Figure 13.1). Model diagrams are a computation techniques for characterizing incomplete knowledge while keying into the knowledge space with functions.

13.2.1 KNOWLEDGE AND PLAN DISCOVERY

Modeling with agent planning is applied where a process to fulfill a task is not predefined, but is created from an intelligent agent. In most cases, the planning

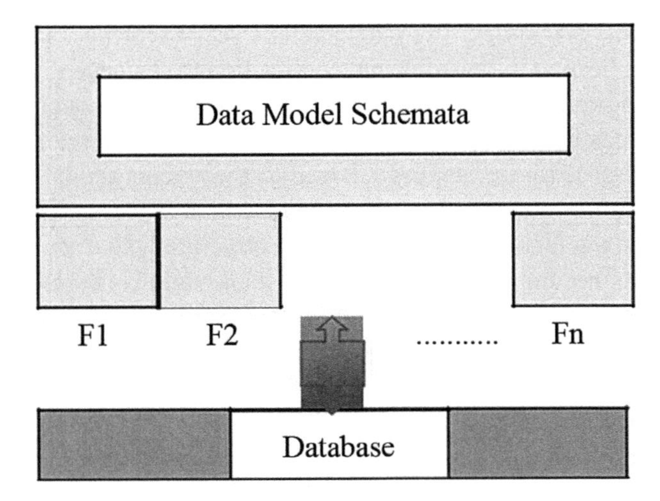

FIGURE 13.1 **(See color insert.)** Keyed database interfaces.

task is additionally complicated by uncertainty in the knowledge about the system, the effect of actors as well as the information amount of sensors.

The incomplete knowledge modeling is treated with knowledge representation on predictive model diagrams. Model discovery at knowledge bases, abbreviated KB's, are realized with specific techniques defined for inference trees. Model diagrams, on the other hand, allow the use of model-theoretically characterizes to handle an incomplete knowledge. Handling incomplete knowledge the approach of this chapter applies generalized predictive diagrams (Nourani, 1995) with specified diagram functions, for example, a search engine, which can be used to search on localized data fields. Data discovery with knowledge represented on a model diagram is viewed as satisfying a goal by getting at data that can instantiates goal satisfiability predicates. With such planning that part of the plan that involves free Skolemized trees is carried along with the proof tree for a plan goal. Computing with diagram functions allows us to key to active visual databases with agents. Diagrams are well-known concepts in mathematical logic and model theory. The diagram of a structure is the set of atomic and negated atomic sentences that are true in that structure. Models for the structures uphold to a deductive closure of the axioms modeled with the applications for rules of inference. Predictive model diagrams (Nourani, 1991–1994; Nourani-Hoppe, 1993) are diagrams in which the elements of the structure are all represented by a specified minimal set of function applicable on nonmonotonic encodings. The language is a set of sentence symbol closed by finite application of negation and conjunction to sentence symbols. Once quantifier logical symbols are added to the language, the language of first-order logic can be defined. A theory T for a language is a set of axioms and rules of inference for the language. Based on the axioms and inference rules, there is a closure on a set of language sentences that the comprises the theory, that is called the theory T. A model for is a structure with a set A. There are structures defined for such that for each constant symbol in the language there corresponds a constant in A. For each function symbol in the language there is a function defined on A; and for each relation symbol in the language there is a relation defined on A. For the algebraic theories we are defining for intelligent tree computing in the forthcoming sections the language is defined from signatures as in the logical language is the language of many-sorted equational logic. The signature defines the language by specifying the function symbols' rarities. The model is a structure defined on a many-sorted algebra consisting of S-indexed sets for S a set of sorts.

13.3 DECISIONS, SPLITTING TREES, AND VECTOR SPANNING MODELS

From the preceding chapters recall that when arranging team playing, there are many permutations on where the players are positioned. Every specific player arrangement is a competitive model. There is a specific arrangement that does best in a specific game. What model is best can be determined with agent player competitive model learning. The decision tree branches compute a Boolean function via agents. The Boolean function is what might satisfy a goal formula on the tree.

The splitting agent decision trees have been developed independently by the first author in ECAI 1994. The agent trees are applied to satisfy goals to attain competitive models for business plans and ERP. The technique is based on Nourani (1996) apply agent splitting tree decisions like what is designed later on the CART system: The ordinary splitting tree decisions are regression-based, developed at Berkeley and Stanford that based on more recent 1996 developments now on CART system Nobel prize worth, deploys a binary recursive partitioning that for our system is applications for the agent AND/OR trees presented in Section 13.2. The term "binary" implies that each group is represented by a "node" in a decision tree, can only be split into two groups. Thus, each node can be split into two child nodes, in which case the original node is called a parent node. The term "recursive" refers to the fact that the binary partitioning process can be applied over and over again. For new directions in forecasting and business planning, e.g., (Nourani, 2002).

13.3.1 MULTIPLAYER DECISION TREES

We have defined specific application areas for multiagent computing to multinational corporations and their strategic management of multinational transactional business models appears in brief at Nourani (1998a). The areas applied to our global planning, external enterprise assessment, and goal-setting applications for operations research and market forecasting (Nourani, 1998b). A specific model starting with a transactional business model is in Nourani (1999a). The organizational knowledge (van Heijst et al., 1994) is one of the main bases to competitive advantage. Enterprise modeling includes stock management, payroll, and advanced administrative tasks by applying for decision support. Figure 13.2 is a glimpse onto applying meas-end analysis decision support where the hidden steps are designed and computed with parameter agents. The obvious planning goal satisfaction applications are where agents apply backward chaining from objectives.

Multiagent planning is based on goal accomplishments with multiplayer processes. Basic multiagent planning examples are Muller and Pischel (1994); and Bazier et al. (1997). In this chapter and the preceding's we are considering goals modeled with a competitive learning problem, with the agents competing on game trees as candidates to satisfy goals hence realizing specific models where the plan goals are satisfied. When a specific agent group "wins" to satisfy a goal the group has presented a model to the specific goal, presumably consistent with an intended world model. For example, if there is a goal to put a spacecraft at a specific planet's orbit, there might be competing agents with alternate micro-plans to accomplish the goal (Figure 13.2).

Therefore, Plan goal selections and objectives are facilitated with competitive agent learning. The intelligent languages (Nourani, 1996,1998) are ways to encode plans with agents and compare models on goal satisfaction to examine and predict via model diagrams why one plan is better than another, or how it could fail. Virtual model planning is treated in the first author's publications where plan comparison can be carried out at VR planning, e.g., (Nourani, 2005).

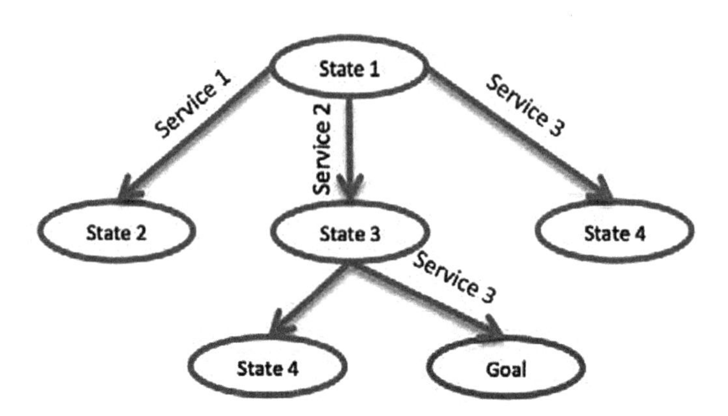

FIGURE 13.2 (See color insert.) Decision Tree on a planning problem for service selection to reach a goal state.

13.4 BIG DATA POSSIBILITIES AND MODELS

Good data governance multiplies the value of big data. Big Data has big value, it also takes organizations big effort to manage well and an effective governance discipline can fulfill its purpose. The Big Data Exponentials: Content, Apps: Consumers have been pledging their love for data visualizations for a while now, and data mining with multimedia discovery is the area being explored. Big data is a popular term used to describe the exponential growth and availability of data, both structured and unstructured. And big data may be as important to business and society as the Internet has become. Why? More data may lead to more accurate analyses. More accurate analyses may lead to more confident decision-making. And better decisions can mean greater operational efficiencies, cost reductions and reduced the risk. There are agent state vectors spanning with models for diagram values that either, true, false, or X-undermined. The cross product on the model diagram vectors is a matrix that can encode a knowledge-base for keying into big data with the X's sparsing the matrix (Nourani, 1992), thus minimizing reaches for bigdata: hence sparse heuristics are possible with functions on sparse model diagrams planning that selects big data segments spanning infinite data. Possible big data encoded on model diagrams with nondeterminism minimizing data segment spans (Section 13.5.3). The diagram is sparse in the sense that there are many X values and on interim computations only minimal sufficient 0,1 values.

13.4.1 AGENT DECISION TREES

Plan goal selections and objectives are based on the attention spans with competitive agent learning. This technique can be also used to solve highly interacting communication problems in a complex web application, for web intelligence. The intelligent languages (Nourani, 1995) are ways to encode plans with agents and compare models on goal satisfaction to examine and predict via model diagrams why one plan is better than another. Modern Reasoning has the ability to simulate commonsense reasoning on certain benchmarks. This is possible because of the integration of background knowledge into the automatically learned facts from sources like Wikipedia. Combining linguistic and AI techniques lead to a vast amount of axioms. These axioms build a semantic graph.

Here a question is transformed into a first-order logic formula and fuzzified via semantic relations in an ontology, for example, Wordnet. Then, background knowledge encoded in knowledge bases like OpenCyc is used to complete the needed facts to do the reasoning for the given question. There are problems to address on how to decide what facts of the background knowledge are relevant (Furbach and Schon, 2016). We have promising options that address such questions through marker passing over the semantic graph. The resulting marked graph selects relevant information out of the millions of facts available to the reasoner. State vector machines-SVM, agent vectors, and data mining referring to the sections above are applied to design cognitive spanning. SVM creates a set of hyperplanes in N-dimensional space that is used for classification, regression, and spanning in our approach. The SVM algorithms create the largest minimum distance to have competitive goals satisfied on models.

The Figure 13.3 depicts selector functions from an abstract view onto a knowledge base and in turn onto a database.

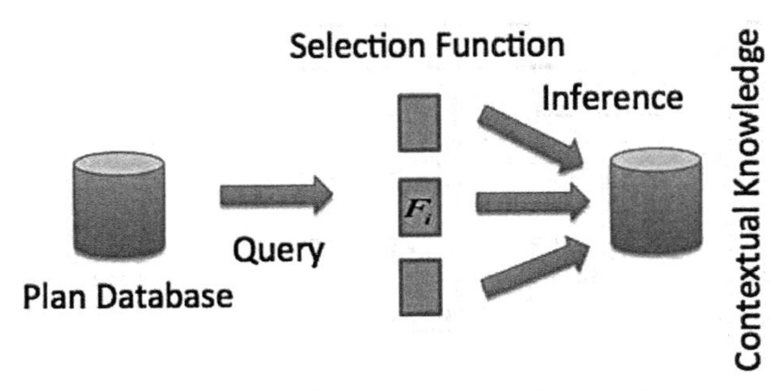

FIGURE 13.3 **(See color insert.)** Inference engine and database inference.

13.4.2 SPARSE BIG DATA

There are agent state vectors spanning with models for diagram values that are, true, false, or X-undermined. A business model diagram can be considered as a basis for a defining a model for an application area, whereby we define a business model with functions with, for example, business objects. The truth-value of describing predicates for the application objects are instantiated with arbitrary functional terms. For agent state vectors we have a matrix with rows a diagram for every spanning agent (Nourani, Lauth, Pedersen, 2014). The matrix is Boolean and sparse coding to big data with the entries that are 0–1-X with X an indeterminate symbol, sparsing the matrix, thus minimizing reaches for big data (Figure 13.4). By applying Boolean matrix multiplications on such sparse matrices, heuristics, etc., computations eventually decide the stages for reaching a heuristic truth-value for every matrix row important for a conclusion. Our newer techniques apply to the emerged a debate about the relative importance of ever bigger data versus ever better predictive techniques to avoid bigger data. Not enough people have the necessary skills to make rigorous use of data. We start by outlining a sparse spanning tree minimizing big data threading State agent vectors F1,...,Fn span a business model viewing section 5.1. The state space is a Boolean-valued matrix on model game trees t1,...tn, for a dynamic state-space agent vectors Matrix is as follows, with each ti being mapped either F,T, or X: the indeterminate symbol, by a computation sequence on the state space vector machine. A computation sequence is F1,F2,....x Fm, with Boolean matrix multiplication. ti term assignment examples are as follows.

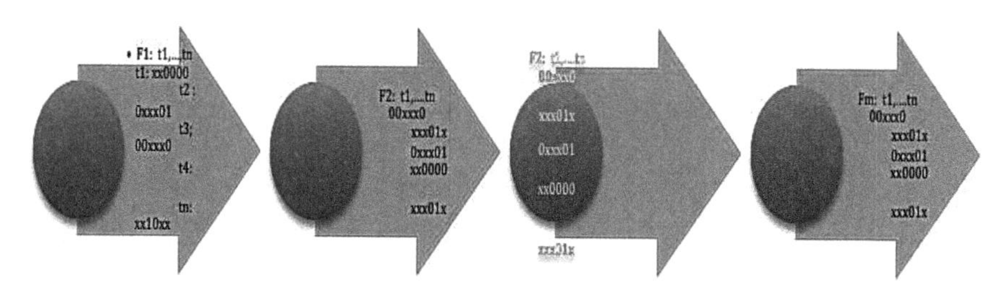

FIGURE 13.4 **(See color insert.)** Sparse nonderministic products.

The sparse matrix determines how the keyed database will minimize the relevant big data sectors. A sparse matrix is a matrix that allows special techniques to take advantage of a large number of "background" (commonly zero or in our case indeterminate X elements). The number of zeros a matrix needs in order to be considered "sparse" depends on the structure of the matrix and the desired operations to perform on it. For example, a randomly generated sparse matrix with entries scattered randomly throughout the matrix is not sparse in the sense of Wilkinson (for direct methods) since it takes time to factor (with high probability and for large enough. The computation sequences on our competitive model techniques are Boolean matrix multiplications on sparse matrices. Applying, for example, Strassen matrix multiplications, we have newer algorithms for sparse minimizing big data areas to reach on a database.

13.4.3 HEURISTICS FOR PREDICTIVE MODELING

First, the author had developed free proof tree techniques since projects at TU Berlin, 1994. The general heuristics to accomplish that is a game tree deductive technique based on computing game tree unfolding projected onto predictive model diagrams. Briefing from preceding chapters with new insights that are essential here, in computer science, specifically in algorithms related to pathfinding, a heuristic function is said to be admissible if it never overestimates the cost of reaching the goal, i.e., the cost it estimates to reach the goal is not higher than the lowest possible cost from the current point in the path.

An optimistic heuristic is used to estimate the cost of reaching the goal state in an informed search algorithm. In order for a heuristic to be admissible to the search problem, the estimated cost must always be lower than or equal to the actual cost of reaching the goal state. The search algorithm uses the admissible heuristic to find an estimated optimal path to the goal state from the current node. An admissible heuristic can be derived from a relaxed version of the problem, or by information from pattern databases that store exact solutions to subproblems of the problem, or by using inductive learning methods. Here, we apply the techniques on goal satisfiability on competitive models briefed on the preceding section. While all consistent heuristics are admissible, not all admissible heuristics are consistent. For tree search problems, if an admissible heuristic is used, the A* search algorithm will never return a suboptimal goal node. The heuristic nomenclature indicates that a heuristic function is called an admissible- heuristic if it never overestimates the cost of reaching the goal, i.e., the cost it estimates to reach the goal is not higher than the lowest possible cost from the current point in the path. An admissible heuristic is also known as an optimistic heuristic (Russell and Norvig, 2002). We shall use the term optimistic heuristic from now on to save ambiguity with admissible sets from mathematics. What is the cost estimate on satisfying a goal on an unfolding projection to model diagrams, for example with SLNDF, to satisfy a goal? Our heuristics are based on satisfying nondeterministic Skolemized trees. The heuristics aim to decrease the unknown assignments on the trees. Since at least one path on the tree must have all assignments defined to T, or F, and at most one such assignment closes the search, the "cost estimate," is no more

13.5 CONCLUSIONS

New bases for splitting decision tree techniques for enterprise business systems planning and design with predictive models are developed. Heuristics on predictive analytics are developed based on ranked game trees from the first authors preceding publications on economic games towards newer applications to decision trees and big data heuristics. Decision systems and game tree applications to analytics towards designing cognitive social media business interfaces are presented. A summary on how big data systems are perceived in the industry was on Data Central 2015: Big data scenarios; I. Thomas, July 2015. Here, are the issues: A Handyman takes servers, a copy of MySQL, towards a basic BI system. An Open Source Guru. A data modeler to take the dynamics of a particular business, product, or process (such as marketing execution) and develop that into a set of data structures that can be used effectively. Deep Divers (often known as Data Scientists). Clarity to balance the inherent uncertainty of the data with the ability to decide on concrete options. A good role is data visualization. Primary data acquisition and privacy. These were the issues that we have addressed in the chapter to a degree. Future research is as follows: a focus on Sections 4.2 and 4.3 areas. For example, algorithms to develop a minimized Boolean matrix product forwarding techniques to complete big data spanning model diagrams. Competitive model learning applications with goal setting bases for decision tree satisfiability is a corollary area to couple with predictive analytics models that accomplish goals on multi-tier business systems. Spanning trees with instantiations that can realize plan goals on matrix products vector state machines coupled with database preprocessor interfaces might have important new applications. Value creation specific application areas are example areas for further research with our techniques.

KEYWORDS

- **big data heuristics**
- **business spanning models**
- **competitive model planning**
- **plan discovery**
- **predictive modeling**
- **sparse spanning agent decision trees**
- **splitting agents decision trees**

REFERENCES

Breiman, L., Friedman, J. H., Olshen, R. A., & Stone, C. J., (1984). *Classification and Regression Trees*. Chapman & Hall (Wadsworth, Inc.): New York.

Furbach, U., & Schon, C., (2016). Commonsense reasoning meets theorem proving. In: Nickles, M., Rovatsos, M., & Weiss, G., (eds.), *Agents*, and *Computational Autonomy* (Vol. 9872, pp. 3–17). Cham: Springer International Publishing. http://doi.org/10.1007/978–3–31945889–2_1

Hendler, J., Tate, A., & Drummond, M., (1990). AI Planning: Systems and Techniques, AI Magazine, Summer.

Le Gall, F., (2012). Rectangular matrix multiplication, *Proceedings of the 53rd Annual IEEE Symposium on Foundations of Computer Science (FOCS 2012)*, pp. 514– 523, arXiv: 1204.1111, doi: 10.1109/FOCS.2012.80.

Liben-Nowell, D., & Kleinberg, J., (2003). The link prediction problem for social networks. *Proceedings of the Twelfth International Conference on Information and Knowledge Management* (pp. 556–559), ACM, New York, NY, USA.

Manyika, J., Chui, M., Brown, B., Bughin, J., Dobbs, R., Roxburgh, C., et al. (2011). *Big Data: The Next Frontier for Innovation, Competition, and Productivity McKinsey Global Institute*. Retrieved from http://www.citeuli ke.org/group/18242/article/9341321.

Muller, J. P., & Pischel, M., (1994). "Modeling interactive agents in dynamic environments." *Proceedings 11th European Conference on AI*, Amsterdam, The Netherlands, John Wiley and Sons. ISBN Nilsson, N.J., "Searching, problem-solving, and game-playing trees for minimal cost solutions." In: Morell, A. J., (ed.), *IFIP 1968* (Vol. 2, pp. 1556–1562). Amsterdam, North-Holland.

Nilsson, N. J., (1971). *Problem Solving Methods in Artificial Intelligence*, New York, McGraw-Hill.

Nilsson, N. J., (1987). *Logical Foundations of Artificial Intelligence,* Morgan-Kaufmann.

Nourani, C. F., (1991). "Planning and plausible reasoning in AI." *Proceedings Scandinavian Conference in AI I* (pp. 150–157). Denmark, IOS Press, Amsterdam.

Nourani, C. F., (1991). Planning and plausible reasoning in artificial intelligence, diagrams, planning, and reasoning. *Proc. Scandinavian Conference on Artificial Intelligence*, Denmark, May, IOS Press.

Nourani, C. F., (1994). Slalom tree computinG-a computing theory for artificial intelligence. *Revised in A.I. Communication, 9*(4), IOS Press.

Nourani, C. F., (1995), *Free Proof Trees and Model-theoretic Planning.* Automated Reasoning AISB, England.

Nourani, C. F., (1996). "Slalom tree computing." Published at AI Communications, *The European Journal on Artificial Intelligence*, *9*(4). IOS Press, Amsterdam, 207–214.

Nourani, C. F., (1997). *Intelligent Tree Computing, Decision Trees and Soft OOP Tree Computing.* Frontiers in soft computing and decision systems papers from the 1997 fall symposium, Boston Technical Report FS–97–04 AAAI ISBN: 1-57735079-0 www.aaai.org/Press/Reports/Symposia/Fall/fs-97-04.html.

Nourani, C. F., (1998). "Intelligent languages-a preliminary syntactic theory." In: Kelemenová, A., (ed.), *Proc. 23rd International Mathematical Foundations Computer Science* (pp. 281–287), Grammatics, Brno, Czech Republic, MFCS'98 satellite workshop on Grammar systems., Silesian University, Faculty of Philosophy and Sciences, Institute of Computer Science, Opava, MFCS SVLN 1450, Springer-Verlag.

Nourani, C. F., (1999a). "Agent computing, KB for intelligent forecasting, and model discovery for knowledge management," In: *The Proceedings AAAI- Workshop on Agent-Based Systems in the Business Context*, Orlando, Florida, AAAI Press.

Nourani, C. F., (2001). The TAIM Intelligent Visual Database 12th International Workshop on Database and Expert Systems Applications (DEXA 2001) 3–7 September 2001 in Munich, Germany. IEEE Computer Society Press ftp.informatik.unitrier.de/~ley/db/conf/dexaw/dexaw2001.html.

Nourani, C. F., (2013). *W-Interfaces, Business Intelligence, and Content Processing.* Invited industry track keynote, IAT-intelligent agent technology, Atlanta.

Nourani, C. F., & Hoppe, T., (1994). *Berlin Logic Colloquium*, Universitat Potasdam.

Nourani, C.F. & Johannes, F., (2015). *A Formal Approach to Agent Planning With Inference Trees*, NAEC, Trieste.

Nourani, C.F. & Johannes, F., (2016). *Decision Trees, Competitive Models, and Big Data Heuristics.* The 4th virtual multidisciplinary conference. www.quaesti.com.

Nourani, C.F. & Oliver, S., (2014). *Multiplayer Games, Competitive Models, Descriptive Computing submitted for the EUROPMENT conferences* (Prague, Czech Republic).

Nourani, C.F., Codrina, L., & Rasmus, U., (2015). Pedersen merging process-oriented ontologies with a cognitive agent, planning, and attention spanning. *Open Journal of Business Model Innovations.*

Russell, S. J., & Norvig, P., (2002). *Artificial Intelligence: A Modern Approach.* Prentice Hall. ISBN 0-13-790395-2.

Thorhildur, J., Michel, A., & Niels, B. A., (2014). In: Birgitta, B. K., & Peter, A. N., (eds.), *Generating Sustainable Value from Open Data in a Sharing Society, IFIP Advances in Information and Communication Technology* (Vol. 429, pp. 62–82). Creating value for all through IT, IFIP WG 8.6 International conference on transfer and diffusion of IT, TDIT, Aalborg, Denmark. ISBN: 978-3-662-43458-1 (Print) 978-3-662-43459-8.

CHAPTER 14

A SEQUENT GAME COMPUTING LOGIC: GAME DESCRIPTIONS AND MODELS

CYRUS F. NOURANI and OLIVER SCHULTE

SFU, Computation Logic Lab, Buranby, BC, Canada

ABSTRACT

Game tree modeling on VMK encodings with a sequent description logic is presented based on computable partition functions on generic game model diagrams. The encoding allows us to reach newer computability areas on game trees based on agent game trees on VMK game models. Novel payoff criteria on game trees and game topologies are obtained. Epistemic accessibility is addressed on game model diagrams payoff computations.

14.1 INTRODUCTION

The chapter presents a sequent description logic model for game tree modeling based on the authors preceding models for von Neumann, Morgenstern, Kuhn games, abbreviated VMK from here on. The computing model is based on a novel competitive learning with agent multiplayer game tree planning. Specific agents transform the models to reach goal plans where goals are satisfied based on competitive game tree learning. Nourani-Schulte (2014, 2015) applied game tree model degrees to encode von Neumann, Morgenstern, Kuhn games that were rendering on a situation calculus on Schulte and Delgrande (2004). The sections outline is as follows. Section 14.2 briefs on knowledge representation on game trees and games as plans, based on an agent AND/OR game tree representation technique. Minimal predictions and predictive diagrams are examined

A preliminary version was published as AI Technical Report Number 45, (1985). Schlumberger Palo Alto Research. In: Franz Baader, et al. (eds.), *The Description Logic Handbook (2003), Theory, Implementation, and Applications*. Aachen University of Technology, ISBN: 9780521781763.

on applications to game trees as goal satisfaction. There is a brief on competitive models as a basis for realizing agent game trees plan goal satisfiability. Agent AND/OR trees are presented to carry on game tree computations. Infinitary multiplayer games are examined and modeling questions with generic diagrams are studied. Section 14.3 presents games, situations, and compatibility with a new descriptive encoding on the game situations that are based on game tuple ordinals computed on generic diagrams. Game descriptions are based on the first author's description logic on generic diagrams (Lipezig, 2010). The techniques are applied to KVM to reach new model-theoretic game tree bases. Section 14.4 briefs on a new KVM encoding from the authors 2012 briefs (e.g., Nourani & Schulte, 2013), where *VMK game function situation, and game tree model diagrams* embed aggregate measures as lifted preorders, towards a sequent description game models.

Section 14.5 presents criteria on a sound and complete game logic based on KVM and hints on applications to compute Nash equilibrium criteria. Furthermore, games and tractability areas are addressed based on descriptive computing, cardinality on concept descriptions, and descriptive computability. Section 14.6 briefs on computing games on proof trees based on plan goal satisfaction, predictions, and game tree models, following the first author's publications past decade.

14.2 COMPETITIVE MODELS AND INFINITARY MULTIPLAYER GAMES

Plan goal selections and objectives are facilitated with competitive agent learning. The intelligent languages (Nourani, 1993d) are ways to encode plans with agents and compare models on goal satisfaction to examine and predict via model diagrams why one plan is better than another or how it could fail.

Games play an important role as a basis to economic theories. Here, the import is brought forth onto decision tree planning with agents. Game tree degree with respect to models is defined and applied to prove soundness and completeness. The game is viewed as a multiplayer game with only perfect information between agent pairs. Upper bounds on determined games are presented. The author had presented a chess-playing basis in 1997 to a computing conference. Game moves are individual tree operations.

14.2.1 INTELLIGENT AND/OR TREES AND SEARCH

AND/OR trees Nilsson (1969) are game trees defined to solve a game from a player's standpoint (Figure 14.1).

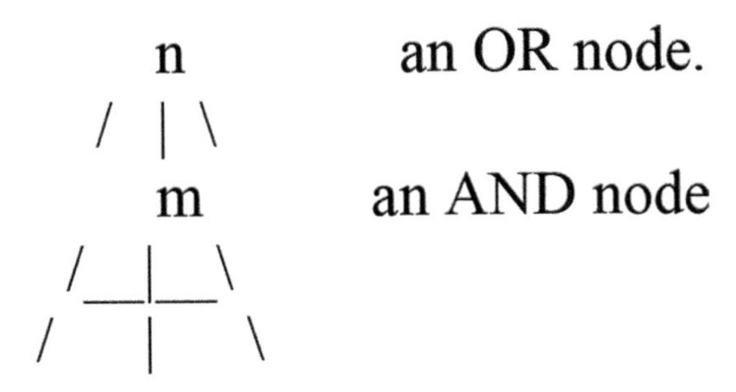

FIGURE 14.1 AND/OR trees.

A solution to the original problem is given by the subgraph of AND/OR graph sufficient to show that the node is solved. A program which can play a theoretically perfect game would have a task like searching and AND/OR tree for a solution to a one-person problem to a two-person game. An intelligent AND/OR tree is where the tree branches are intelligent trees.

The branches compute a Boolean function via agents. The Boolean function is what might satisfy a goal formula on the tree.

An intelligent AND/OR tree is solved if *f* the corresponding Boolean functions solve the AND/OR trees named by intelligent functions on the trees. Thus, node m might be f(a1,a2,a3) & g(b1,b2), where *f* and *g* are Boolean functions of three and two variables, respectively, and ai's and bi's are Boolean-valued agents satisfying goal formulas for *f* and *g* (Figure 14.2).

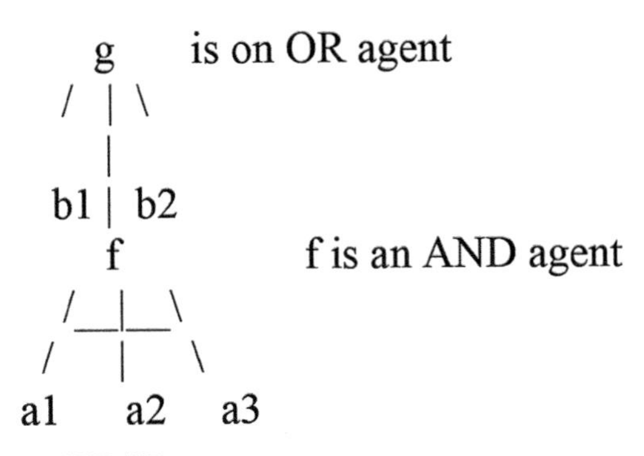

FIGURE 14.2 Agent AND/OR trees.

The chess game trees can be defined by agent augmenting AND/OR trees (Nourani, 1990s). For the intelligent game trees and the problem-solving techniques defined, the same model can be applied to the game trees in the sense of two-person games and to the state space from the single agent view. The two-person game tree is obtained from the intelligent tree model, as is the state space tree for agents. To obtain the two-person game tree the cross-board-coboard agent computation is depicted on a tree. Whereas the state-space trees for each agent are determined by the computation sequence on its side of the board–coboard.

A **game tree degree** is the game state a tree is at with respect to a model truth assignment, e.g., to the parameters to the Boolean functions above. Let generic diagram or G-diagrams be diagrams definable by specific functions. Intelligent signatures (Nourani, 1996a) are signatures with designated multiplayer game tree function symbols. A soundness and completeness theorem is proved on the intelligent signature language (Nourani, 1999a). The techniques allowed us to present a novel model-theoretic basis to game trees, and generally to the new intelligent game trees. The following specifics are from Nourani (1999).

14.2.2 PREDICTIONS AND DISCOVERY

Minimal prediction is an artificial intelligence technique defined since the author's model-theoretic planning project. It is a cumulative nonmonotonic approximation (Nourani, 1999c) attained with completing model diagrams on what might be true in a model or knowledge base. A *predictive diagram* for a theory T is a diagram D (M), where M is a model for T, and for any formula q in M, either the function f: q → {0,1} is defined, or there exists a formula p in D(M), such that T U {p} proves q; or that T proves q by minimal prediction. A *generalized predictive diagram* is a predictive diagram with D (M) defined from a minimal set of functions. The predictive diagram could be minimally represented by a set of functions {f1,...,fn} that inductively define the model. The free trees we had defined by the notion of probability implied by the definition, could consist of some extra Skolem functions {g1..., gk} that appear at free trees. The *f* terms and *g* terms, tree congruences, and predictive diagrams then characterize partial deduction with free trees. The predictive diagrams are applied to discover models to the intelligent game trees. Prediction is applied to plan goal satisfiability and can be combined with plausibility (Nourani, 1991) probabilities, and fuzzy logic to obtain, for example, confidence intervals.

Games, Situations, and Compatibility. Now let us examine the definition of the situation and view it in the present formulation.

Definition 2.1: A situation consists of a nonempty set D, the domain of the situation, and two mappings: *g* and *h*. *g* is a mapping of function letters into functions over the domain as in standard model theory. h maps each predicate letter, pn, to a function from Dn to a subset of {t,f}, to determine the truth value of atomic formulas as defined below. The logic has four truth values: the set of subsets of {t,f}.{{t},{f},{t,f},0}. The latter two is corresponding to inconsistency, and lack of knowledge of whether it is true or false. □

Due to the above truth values, the number of situations exceeds the number of possible worlds. The possible worlds being those situations with no missing information and no contradictions. From the above definitions, the mapping of terms and predicate models extend as in standard model theory. Next, a **compatible set of situations** is a set of situations with the same domain and the same mapping of function letters to functions. In other words, the situations in a compatible set of situations differ only on the truth conditions they assign to predicate letters.

Definition 2.2: Let M be a structure for a language L, call a subset X of M a generating set for M if no proper substructure of M contains X, i.e., if Mis the closure of X U {c[M]: c is a constant symbol of L}. An assignment of constants to M is a pair <A,G>, where A is an infinite set of constant symbols in L and G: A Æ M, such that {G[a]: a in A} is a set of generators for M. Interpreting a by g[a], every element of M is denoted by at least one closed term of L[A]. For a fixed assignment <A, G> of constants to M, the diagram *of M, D <A, G>[M]* is the set of basic [atomic and negated atomic] sentences of L[A] true in M. [Note that L[A] is L enriched with set A of constant symbols.]

Generic diagrams, are diagrams for models defined by a specific function set, for example, Σ1 Skolem functions.

Definition 2.3: A Generic diagram for a structure M is a diagram D<A,G>, such that the *G* in definition 2.2 has a proper definition by specific function symbols.

The dynamics of epistemic states as formulated by generalized diagrams (Nourani 1991, 1998) is exactly what addresses the compatibility of situations. What it takes to have an algebra and model theory of epistemic states, as defined by the generalized diagram of possible worlds is exactly what (Nourani 1991, 1998) leads us to.

Theorem 1 (Nourani, 1994): Two situations are compatible if*f* their corresponding generalized diagrams are compatible with respect to the Boolean structure of the set to which formulas are mapped (by the function h above, defining situations).

Proof: The G-diagrams, definition 3.3, encode possible worlds and since we can define a one-one correspondence between possible worlds and truth sets for situations, computability is definable by the G-diagrams.□

The computational reducibility areas were briefed at Nourani (1998, 2009).

14.2.3 NEW MODELS MATHEMATICAL BASIS

Let us start from certain model-theoretic premises with propositions 3.1 and 3.2 known form a basic model theory.

Proposition 2.1: Let R and B be models for L. Then, R is isomorphically embedded in B if B can be expanded to a model of the diagram of R.

Proposition 2.2: Let R and B be models for L. Then, R is homomorphically embedded in B if B can be expanded to a model of the positive diagram of R.

Definition 2.4: Given models R and B, with generic diagrams DR and DB we say that DR homomorphically extends DB if there is a homomorphic embedding f: A → B.

14.2.4 COMPATIBILITY ORDERING ON INCREMENTAL GAME TREES

Agent game tree generic diagram satisfiability based on the above compatibility criteria on game tree nodes implies the following.

Proposition 2.3: A game tree node generic diagram player at an agent game tree node is satisfiable if the subtree agent game trees situations have a compatible generic diagram to the node.

Proposition 2.4: Agent game tree generic diagrams encode compatibility satisfying the von Neumann, Morgenstern, Kuhn game tree criteria for the node player.

Theorem 4: A game tree is solved at a node if there is an elementary embedding on the generic diagrams, sequentially upward to the node, corresponding to VMK played at the node (Nourani, 2015).

14.3 DESCRIPTIONS AND GAMES

Von Neumann and Morgenstern VM-Short designed game theory as a very general tool for modeling agent interactions. VM game theory in a logical formalism, so as to enable computational agents to employ game-theoretical models and reasoning for the interactions that they engage in. Let us consider VM games in which the agents can take at most countably many actions and in which the agents' payoff functions are continuous. (Schulte et al., 2004) show that every such VM game *G* has an axiomatization Axioms(G) in the situation calculus that represents the game in the following strong sense:

The game *G* itself is a model of Axioms (G), and all models of Axioms (G) are isomorphic. It follows that the axiomatization is correct, in the sense that Axioms (G) entails only true assertions about the game G, and complete in the sense that Axioms (G) entails all true assertions about the game G.

Let us call axiomatizations A(G) where all models of A(G) are isomorphic a categorical axiomatization.

Definition 3.1: (von Neumann, Morgenstern, Kuhn). A *sequential game G* is a tuple ⟨N, H, player, fc,{Ii},{ui}⟩whose components are as follows.

1. A finite set N (the set of *players).*
2. A set H of sequences satisfying the following three properties.
   (a) The empty sequence ∅ is a member of H.
   (b) If σ is in H, then every initial segment of σ is in H.
   (c) If h is an infinite sequence such that every finite initial segment of h is in H, then h is in H.

   Each member of H is a *history;* each element of a history is an *action* taken by a player. A history σ is *terminal* if there is no histories ∈H, such that σ ⊆ s. (Thus all infinite histories are terminal.) The set of terminal histories is denoted by Z. The set of actions available at a finite history s is denoted by A(s) = {a: s⊆a∈H}.
   A function *player* that assigns to each nonterminal history a member of N⊆ {c}. The function *player* determines which player takes an action after the history s. If the player(s) = c, then it is "nature's" turn to make a chance move.
3. A function fc that assigns to every history s for which player(s) = c a probability measure fc(·|s) on A(s). Each probability measure is

independent of every other such measure. (Thus fc(a|s) is the probability that "nature" chooses action a after the history.)

4. For each player i∈N an *information partition* Ii defined on{s∈ H: player(s)=i}. An element Ii of Ii is called an *information set* of Player i. We require that if s, s′ are members of the same information set Ii, then A(s) =A(s′).
5. For each player i∈ N a *payoff function* ui: Z → R that assigns a real number to each terminal history.

An element *Ii* of *Ii* is called an *information set* of Player *i.* We require that if *s, s* are members of the same information set *Ii,* then *A(s)* = *A(s).* 6. For each player *i*∈ N a *payoff function ui*: Z → R that assigns a real number to each terminal history.

To examine game models we present an algebraic characterization that is not specific to the game so far as the probability measure assignment on game sequences are concerned. The information set on the game nodes are encoded on the game diagram. Whatever the game degree is at the node is how the following sequence is carried on. On that basis when can examine games based on satisfying a goal, where the compatibilist is characterized on specific VMK model embedding that is stated in the following section.

Let us state a categorical VMK situation. For the time being, consider modeling zero-sum games and not adversarial games not to confuse the modeling questions. Modeling agent games based on game tree satisfiability does not require explicit probability measures. The information partition can be encoded such that whatever the unknown probabilities are, that in any case are very difficult to ascertain, the game tree node moves can be determined as the game develops. In Section 14.3.7 we present a treatment on measures and payoffs.

Definition 3.2: A ***VMK game function situation*** consists of a domain *N x H, where N is the natural numbers, H a set of game history sequences, and a mapping pair g,f. f maps function letters to (agent) functions and g maps pairs from* <*N,H*> *to{t.f}.*

Notes: On condition 5 at VMK definition 4.4 the information partition for a player I is determined at each game node by the game diagram game tree degree (Section 14.4.1) at that node that is known to that player, c.f. (Nourani, 1994, 1996).

The basic intricacy is what you can say about conditions on H on the encoding on a basic definition: H for example: that agent player plays to increases the pay on ordinals on every move. That is, there is an ordering on the functions on <N,H>, that is from a mathematical absolute. That is essentially a pre-ordered mapping to {t,f} where the ordering is completed on stages. However, the ordering is not known before a game goal is satisfied.

14.3.1 COMPATIBILITY ORDERING AND VMK GAME TREES

Agent game tree generic diagram satisfiability based on the above compatibility criteria on game tree nodes implies the following.

Definition 3.3: Two generic diagrams D1 and D2 on the same logical language £ are compatible if forever assignment to free variables to models on £, every model M for £, M $\models$ D1iff M $\models$ D2.

Proposition 3.1: A game tree node generic diagram on a player at an agent game tree node is satisfiable if the subtree agent game trees situations have a compatible generic diagram to the node.

Proof: Follows from game tress node satisfiability, definitions 3.1, 4.4, and 3.3

Proposition 3.2: Agent game tree generic diagrams encode compatibility on satisfying the VMK game tree criteria for the node player.

Proof: Follows from VMK definitions and proposition 4.1.

Proposition 3.3: The information partition for each player I at a node is Ṿ D<f_I> where D<f_I> is the diagram definable to the player I's agent functions at that node, where Ṿ is the infinite conjunction.

Proof: Since VMK has countable agent assignments at each game tree node there is a diagram definable on the partition where countable conjunction can determine the next move.

For example, viewing the diagrams on a Tableaux's (Nourani, 2000), a conjunction indicates what the information partition state is for an agent.

Theorem 2: A game tree is solved at a node if there is an elementary embedding on the generic diagrams, sequentially upward to the node, corresponding to VMK played at the node, that solves the root node.

Proof: Elementary embedding has upward compatibility on tree nodes as a premise. Proposition 3.4 grants us that a VMK node check is encoded on diagram compatibility. Applying Proposition 2.1 we can base diagram check the ordinal assignments in isomorphic embeddings to compute a countable conjunction, hence an ordinal. On proposition, we have that compatibility is definable on a countable conjunction over the agent partition diagrams. Therefore, at the root node, a game goal formula φ is solved, if at each game tree node to the terminal node the $D<f_I>$ on the nodes, for every model Mj, and a game sequence <S>, Msj |= $D<f_I>$ iff Msj–1 |= $D<f_I>$ on the ordered sequence s ε <H x I>, j an ordinal <|S|.

First, the author characterizes agent morphisms, in for example (Nourani, 2005), to compute agent state-space model computations. So we can, for example, state the following.

From the preceding, we can address on newer publications how model epistemic can be treated. The relationship to the diagram is an information set, an agent's knowledge is characterized by the diagram. What we have not stated is whether we can use modal logic or not. For starters, the diagram is no longer finite unless you make further assumptions because you have to model what agents know and believe about each other. The above proposition allows us to treat infinite conditions. Model diagrams were treated in the first author's publications, for example, Nourani (2000).

14.4 SEQUENT DESCRIPTION MODELS

The first author's 1996 publications on Slalom Tree computing presents canonical tree computing with agent splitting decision trees. A brief was at ECAI 1994. The logical areas are stated in brief here. For having a comprehensive treatment of agent decision trees.

Definition 4.1: A function *f* is **accessible** by an agent if either *f* or the agent function symbol are defined as proper language signature terms. A subtree t has access to an agent *g* if there is a function symbol on t accessible by g. □

Definition 4.2: The tree intelligence degree, TID, is defined by induction on tree structures:

(i) a constant function symbol *f* has TIDf;
(ii) for a string function f, and tree f(t1…,tn) the TID is defined by U TID (ti: : f), where (ti: : f) refers to a subtree of ti with access to *f* via an agent function symbols on ti;
(iii) for a splurge function f, TID is defined by U TID (f: ti), where f: ti is the updated tree ti since a single access to ti from f.□

Let us commence with the preliminary logical basis as far as proofs and models.

Definition 4.3: Let us say that a logical theory T on agent intelligent syntax is a sequent descriptive intelligent-SDI theory iff

(i) Every proof step preserves soundness and completeness with respect to the information partition function game tree models.
(ii) All proof rules are sound with respect to the description logic language, e.g., VMK descriptions.

We state T<SDI> |- φ where T is an SDI theory.

Theorem 3: Let T be a SDI theory. Then, T is

(a) A Sound logical theory iff every axiom or proof rule in T is TID preserving and every proof step is sound with respect to the agent proof tree information partition models.
(b) A complete logical theory iff there is a generic diagram G, where the assignment of constants to M is a pair <A,G>, where A is an infinite set of constant symbols and G: A → M, such that {G[a]: a in A} the set of generators for M with *G* definable with the generic diagram, where M is a structure for the intelligent syntax language on which T is defined.

Proof: Follows from Nourani (1994).

14.5 COMPETITIVE MODELS AND SEQUENT DESCRIPTIONS

On a forthcoming publication (Nourani & Schulte, 2012) with a mathematics glimpse to appear at Nourani & Schulte (2013), the authors present explore agent game plan competitive model computing. Agent game trees and agent intelligent languages are presented with a description computing logic models. A basis to model discovery and prediction planning is stated. The techniques are developed on a descriptive game logic where model compatibility is characterized on von Neumann, Morgenstern, Kuhn, abbreviated VMK, game descriptions model embeddings, and game goal satisfiability. Game tree nodes information partition functions allow us to embed measures on game trees where agent sequence actions, be that random sequences are modeled. The import is that game tree modeling, KVM encodings based on partition functions on generic game model diagrams, allows us to reach newer computability areas on game trees based on agent game trees on KVM game models. Novel payoff criteria on game trees and game topologies are obtained.

Definition 5.1: Two generic diagrams D1 and D2 on the same logical language £ are compatible iff forever assignment to free variables to models on £, every model M for £, $M \models D1$ iff $M \models D2$.

Proposition 5.1: A game tree node generic diagram on a player at an agent game tree node is satisfiable iff the subtree agent game trees situations have a compatible generic diagram to the node.

Proof: Follows from game tress node satisfiability, and definitions.

Proposition 5.2: Agent game tree generic diagrams encode compatibility on satisfying the VMK game tree criteria for the node player.

Proof: Follows from VMK definitions and proposition 5.1.

Proposition 5.3: The information partition for each player I at a node is Ṿ $D<f_I>$ where $D<f_I>$ is the diagram definable to the player I's agent functions at that node, where Ṿ is the infinite conjunction.

Proof: Since VMK has countable agent assignments at each game tree node there is a diagram definable on the partition where countable conjunction can determine the next move.

From Definitions 3.1 and 3.2 and the preceding section, we define a sequent game description modes as follows.

Definition 5.2: *A sequent game description model* is a structure that assigns a function to agent function symbols, relations to every predicate letter, and assigns function compositions to every game action sequence consistent with Axioms (G). Furthermore, the sequent description structure models the categorical game axioms Axioms (G) such that Axioms (G) entails only true assertions about the game G, and complete in the sense that Axioms (G) entails all true assertions about the game G.

Theorem 4: Let T be a SDI theory. Then, T is:

(a) A sound logical theory iff every axiom or proof rule in T is TID preserving and every proof step is sound with respect to the sequent description model partitions.
(b) A complete logical theory iff there is a generic diagram *G* where the assignment of constants to M is a pair <A,G>, where A is an infinite

set of constant symbols and G: A → M, such that {G[a]: a in A} the set of generators for M with *G* definable with the generic diagram, where M is a structure for the intelligent syntax language on which T is defined. M is a sequential description structure.

Proof: Follows from Nourani (1994) and chapter preceding by the authors since Nourani (1996, 2015)

14.6 CONCLUSIONS

The techniques apply to both zero-sum and arbitrary games. The new encoding with a *VMK game function situation,* where agent sequence actions, are have embedded measures. The import is that game tree modeling on VMK encodings are based on computable partition functions on generic game model diagrams. The encoding allows us to reach newer computability areas on game trees based on agent game trees on VMK game description models. Novel payoff criteria on game trees and game topologies are obtained. Epistemic accessibility is addressed on game model diagrams payoff computations. Furthermore, criteria are presented to reach a sound and complete game logic based on VMK and hints on applications to compute Nash equilibrium criteria and a precise mathematical basis is stated.

KEYWORDS

- **VMK encodings**
- **VMK game description models**
- **VMK game function situation**

REFERENCES

Brazier, F. M. T., Dunin-Keplicz, B., Jennings, N. R., & Treur, J., (1997). DESIRE: Modeling multi-agent systems in a compositional formal framework, In: Huhns, M., & Singh, M., (eds.), *International Journal of Cooperative Information Systems* (Vol. 1). Special issue on formal methods in cooperative information systems.

Gale, D., & Stewart, F. M., (1953). "Infinite games with perfect information." In: *Contributions to the Theory of Games, Annals of Mathematical Studies* (Vol. 28). Princeton.
Genesereth, M. R., & Nilsson, N. J., (1987). *Logical Foundations of Artificial Intelligence*. Morgan- Kaufmann.
Hagel, J. V., & Armstrong, A. G., (1997). *Net Gain-Expanding Market Through Virtual Communities*. Harvard Business School Press, Boston, Mass.
Hendler, J., Tate, A., & Drummond, M., (1990). *AI Planning: Systems and Techniques*. AI magazine, Summer.
Muller, J. P., & Pischel, M., (1994). "Modeling interactive agents in dynamic environments." *Proceedings 11th European Conference on AI,* Amsterdam, The Netherlands, John Wiley and Sons. ISBN 0471950696.
Nilsson, N. J., (1971). *Problem Solving Methods in Artificial Intelligence*, New York, McGraw-Hill.
North-Holland, & Nilsson, N. J., (1969)."Searching, problem-solving, and game-playing trees for minimal cost solutions." In: Morell, A. J., (ed.), *IFIP 1968* (Vol. 2, pp. 1556–1562). Amsterdam, North-Holland.
Nourani, C. F., (1983). "Equational intensity, initial models, and AI reasoning," Technical report: A conceptual overview. In: *Proceedings Sixth European Conference in Artificial Intelligence*. Pisa, Italy.
Nourani, C. F., (1991). "Planning and plausible reasoning in AI." *Proceedings Scandinavian Conference in AI, May,* Denmark, 150–157, IOS Press.
Nourani, C. F., (1993d). "Intelligent languages- a preliminary syntactic theory." In: Kelemenová, A., (ed.), *Proceedings of the Mathematical Foundations CS'98 Satellite Workshop on Grammar Systems*. Silesian University, Faculty of philosophy and sciences, Institute of Computer Science, Opava, pp. 281–287.
Nourani, C. F., (1994c). "*A Theory for Programming With Intelligent Syntax Trees and Intelligent Decisive Agents-Preliminary Report*." 11th European Conference A.I., ECAI Workshop on DAI Applications to Decision Theory, Amsterdam.
Nourani, C. F., (1995). "*Free Proof Trees and Model-theoretic Planning*." Automated Reasoning AISB, England.
Nourani, C. F., (1996a). "Slalom tree computing," AI communications, *The European AI Journal*, IOS Press.
Nourani, C., (1997). *Intelligent Tree Computing, Decision Trees and Soft OOP Tree Computing*. Frontiers in soft computing and decision systems papers from the 1997 fall symposium, Boston Technical Report FS–97–04AAAI.
Nourani, C. F., (1997). "Multiagent chess games." *AAAI Chess Track, Providence*, RI.
Nourani, C. F., (1999a). "Agent computing, KB for intelligent forecasting, and model discovery for knowledge management." *AAAI-Workshop on Agent-Based Systems in the Business Context,* Orlando, Florida.
Nourani, C. F., (1999b). "Virtual tree planning." *Abstract accepted at the East Coast Algebra Conference*. North Carolina, Abstract recorded at the author's name, http://www.logic.univie.ac.at.
Nourani, C. F., (1999c). *Infinitary Multiplayer Games*. Summer logic colloquium, Utrecht.
Nourani, C. F., (1999d). *Nonmonotonic Model Culmination*. ASL, UCSD. Proceedings BSL.
Nourani, C. F., (2009). A descriptive computing, information forum, Leipzig, Germany. SIWN2009 Program, 2009. *The Foresight Academy of Technology Press International Transactions on Systems Science and Applications* (Vol. 5, No. 1, pp. 60–69).

Nourani, C. F., & Schulte, O. (2012). *Multiagent Games, Competitive Models, and Descriptive Computing*. SFU, Vancouver, Canada, Preliminary Draft outline, New SFU Computational Logic Lab, Technical Report, Vancouver, Canada.

Nourani, C. F., & Schulte, O. (2013). "Competitive models, descriptive computing, and Nash games," SFU, Burnaby Canada, AMS-Oxford, Mississippi.

Patel-Schneider, P. F., (1990). A decidable first-order logic for knowledge representation. *Journal of Automated Reasoning, 6*, 361–388.

Schulte, O., & Delgrande, J., (2004). Representing von Neumann-Morgenstern games in the situation calculus. *Annals of Mathematics and Artificial Intelligence, 42*(1–3), 73–101. (Special issue on multi-agent systems and computational logic). A shorter version of this paper appeared in the 2002 AAAI workshop on decision and game theory.

Schulte, O., (2003). Iterated backward inference: An algorithm for proper rationalizability. *Proceedings of TARK IX (Theoretical Aspects of Reasoning About Knowledge)*, Bloomington, Indiana, pp. 15–28. ACM, New York. Expanded version with full proofs.

The Knowledge Engineering Review, (2012). Special Issue 02 (Agent-Based Computational Economics) (Vol. 27, No. 02). Cambridge Online Journals.

Wilkins, D., (1984). "Domain independent planning: Representation and plan generation." *AI, 22*(3), 269–301.

INDEX

D

L

M

Q

R

T

U

V

W

Y

Z

9781771887298